P9-CBL-490

Illuminations

A BESTIARY

ROSAMOND WOLFF PURCELL

STEPHEN JAY GOULD

W · W · N O R T O N & C O M P A N Y
N E W Y O R K · L O N D O N

Published simultaneously in Canada by Penguin Books Canada Ltd, 2801 John Street, Markham, Ontario L3R 1B4

The text of this book is composed in Garamond no. 3, with display type set in Zaph Chancery Lite. Composition by the Maple-Vail Book Manufacturing Group
Printed and bound by Dai Nippon Printing Co., Ltd., Tokyo, Japan.
Book design by Antonina Krass

First Edition

Library of Congress Cataloging-in-Publication Data
Purcell, Rosamond Wolff.
Illuminations : a bestiary.
1. Zoological specimens—Pictorial works. 2. Zoology—Miscellanea—Pictorial works. I. Gould, Stephen Jay. II. Title.
QL46.P87 1986 779'.32 86-8713

ISBN 0-393-02374-5

W. W. Norton & Company, Inc., 500 Fifth Avenue, New York, N.Y. 10110
W. W. Norton & Company Ltd., 37 Great Russell Street, London WC1B 3NU

1 2 3 4 5 6 7 8 9 0

Dedicated to the Museum of Comparative Zoology,
Harvard University

Contents

Introduction

Death strips information from an organism, layer by layer. Dry bones form our standard metaphor for the end of this process, though it goes further to triturated dust. We view a horse's jaw as discarded residue, though tradition records an occasional use, notably by Samson (asses, to taxonomists, are horses, not degraded stepsons of nobility).

Sometime during the mid-seventeenth century, in northern Europe, a young oak sapling encountered a horse's jaw and, instead of pushing it aside in upward growth, grew through and around, incorporating the jaws into its living substance—a fine symbol for the quick and the dead, the fusion of unlikes. In 1649, King Frederick III gave this curiosity to the Danish naturalist, Ole Worm. Worm figured and described the strange jaw in the catalog of his "cabinet," or private museum—*Equina mandibula inferior, trunco quercino ita innata, ut insertionis*

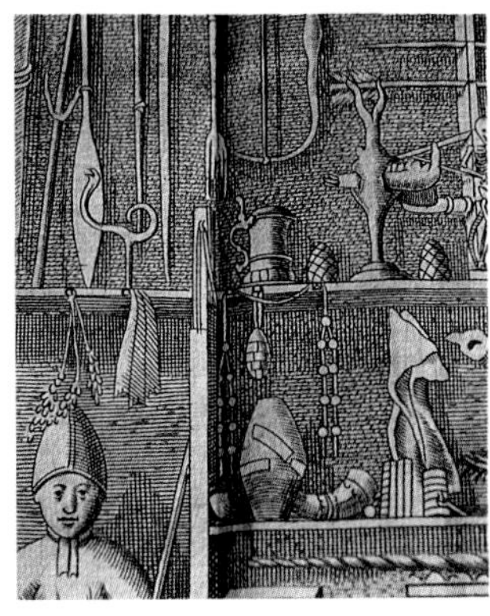

nulla appareant vestigia . . . (the lower jaw of a horse, so joined to an oak branch that no traces of its insertion remain). More importantly for our story, he included this specimen as an inconspicuous dot high on a shelf in his wonder room—surely the most famous and familiar illustration of a naturalist's cabinet from this age of curiosities.

I had stared long and hard at this celebrated figure, but never noticed the horse's jaw; I wonder if anyone ever has. In 1985, Rosamond Purcell found

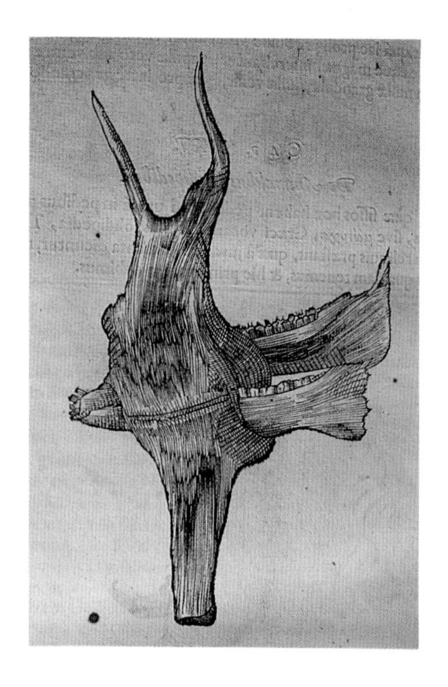

Worm's original specimen in the Zoology Department of the University of Copenhagen. For 300 years, we have known this jaw only in the two dimensions that art presented. But now, taken from its shelf, we can see the other side—a small point perhaps, but for me comparable, in symbolic worth, to our first photographs of the moon's invisible side. (And I, for one, never experienced a

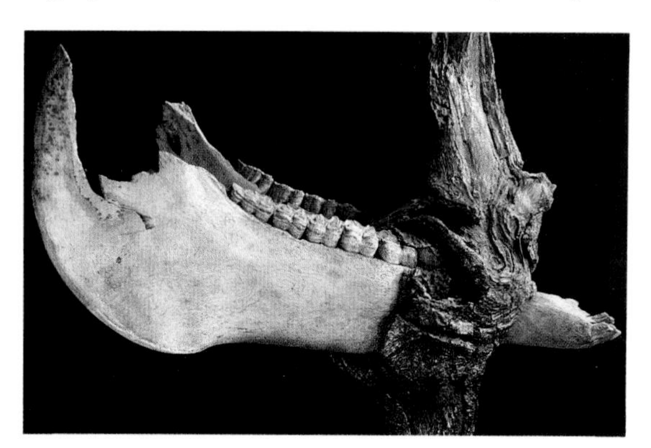

greater moment of pure intellectual joy than my first view of this obverse, previously hidden from the sight of absolutely every human being, from the first australopithecine on the savannas of Africa to Darwin, Einstein, and all the great dreamers of history.) Art, at its best, can add a dimension to our perceptions.

This book treats our relationship to the natural world in a novel way. It comments upon life from a perspective that could not be more different from the traditional photographic essay in natural history, though it confutes or contradicts nothing about the ordinary mode. The standard tradition in natural history tries to capture organisms as they "are" in nature—and it favors the beautiful and pleasant, rather than the disturbing or ugly. Its conceit is factuality and separation from human concerns—nature as she is, away from the corruptions of civilization. Its twin claims are *objectivity* and *beauty*.

This is not a book in natural history, so conceived. It is about a particular mode of *interaction* with organisms, a path known only to professionals, and therefore all the more instructive because hidden from public view. This book gathers its material from the back rooms of museums; it captures attitudes expressed in the way we store and collect organisms that are *not* on public display. This book is about information lost in partial preservation, and dimensions added by treatment that exposes, in ways all the more telling because unconscious, our intimate connections with all life.

That old cliché about the tip of an iceberg applies here with special force. The displays of our museums include little of the totality in drawers and cabinets behind locked doors or on upper floors. And what they do contain has usually been dressed up to match the twin conceits of traditional natural history—beauty and fidelity to an objective nature unconcerned with human presences. The invisible specimens of the back rooms are our finest sources of art and insight—our most complex objects, intertwined with our thoughts and fears. The back rooms now hold Ole Worm's peculiar jaw.

Most specimens in this book came from the collections of the Museum of Comparative Zoology at Harvard University. But the back rooms are an invisible world, and Rosamond has scoured them from Paris to Leiden in order to muse upon this hidden community.

I believe that organisms have, through their evolution, an irreducible and inalienable status in and for themselves. But we can only speak of them in terms of their meaning for us; culture and mind permeate our world of discourse. Our traditional uses have been exploitive and often tragic—from outright carnage at the worst to viewing animals only as sources of moral messages at the mildest. The classic tradition of natural history tries to undo these sadnesses by proclaiming a divorce and seeking objective representation. But why not admit that we cannot see natural objects except in our terms, and struggle instead to make the union instructive. Why not seek out the hidden objects on shelves of the world's great museums, and ask what they convey, or explore what we might say to them.

These photographs speak of layers of data lost, and meaning added—a mean-

lower portion of lateral line
on the left side, the second por
five scales beyond the und
four scales beyond.

ing that records interaction based upon our own limitations, not an emanation from the animal itself. Thus, several photographs present a kind of optical illusion born of expectations that force us to see parts of animals for what they are not: a fossil eurypterid becomes a New England gravestone; the wing tips of a moth are cartoon faces of morning worms; a view through the skull of a howler monkey becomes an old man with hunched shoulders; a face smiles from a turtle's spine, its "mouth" the junction between two vertebrae. Others fool us because, in the absence of a scale, we read a shape as something much larger or smaller in *our* world. Whale bones become landscapes, and the cusps of a mastodon tooth a range of mountains (with clouds from cotton that cradles the tooth in its museum box).

One tradition proclaims that objects of art should speak for themselves, and that commentary (particularly from someone else) can only clutter, or become an unwelcome intrusion of an alien ego. In a stronger version, which we heartily reject, art and science (sometimes abstracted to a false and silly antithesis between feeling and intellect) should follow their separate and legitimate pathways. From the beginning, we thought that this book needed text. The photographs were made with this experiment in mind; their full beauty requires some technical information about what parts we see, of what creatures, in what settings. The theme of human uses and perceptions can only be played off against a knowledge of taxonomy and evolutionary history. This book then becomes, in part, our contribution to the healing of a false and dangerous dichotomy in human knowledge. I have always wanted to play W.S. Gilbert—who maintained, after all that "Darwinian man, though well behaved, at best is only a monkey shaved"—to a more sublime Sullivan.

Taxonomy, or the science of classification, is the most underrated of all disciplines. Dismissed by the uninformed as philately gussied up with jargon, classification is truly the mirror of our thoughts, its changes through time the best guide to the history of human perceptions. The surest sign of our intent must therefore be the taxonomic scheme of our photographic arrangement. We have

chosen an old tradition, the bestiary arranged by letters of our alphabet.

To some, this may seem a disappointing and unworthy retreat from modern enlightenment. What, after all, can the arbitrary names applied by people in a fortuitous sequence of letters say about the "real" status of organisms and the messages of their lives. This choice may seem even worse when we delve into medieval bestiaries and immerse ourselves in the neo-Platonic perspective that granted names as much meaning as things, and therefore did not deem it unthinkable that the proper abracadabra might transmute mercury to gold. In the old bestiaries, no name could be an accident (however forced the derivation of meaning), and each animal existed only for our moral instruction. Thus, we learn that *capra* (the goat) is so named "because she strives to attain the mountain crags (aspera *cap*tet)." And so "Our Lord Jesus Christ is partial to high mountains" because we read in the *Song of Songs:* "Behold my cousin cometh like a he-goat leaping on the mountains." Moreover, a goat's vision is acute, "and they see everything and know men from afar, this symbolizes our Lord, who is the Lord

God of all knowing" (all quotes from T.H. White's *The Bestiary,* Capricorn books [another goat allusion], 1960—translated from a twelfth-century Latin bestiary).

While we acknowledge these superseded features of classical bestiaries, we choose this format with active pleasure for several reasons. First, in continuity with tradition, it captures forthrightly the conviction that we can only see organisms in their relation with us. Second, it avoids other standard taxonomic devices that do embody what we seek most heartily to avoid—judgments of worth in human terms (the ladder of simple to complex, or time of evolutionary origin relative to our late arrival). Thus, we hope that the honest acknowledgment of intrinsic relationship will place the necessity of ties up front, thereby forcing us to contemplate explicitly what the nature of those links should be. Finally, the very arbitrariness of arrangement by letters best expresses our deepest conviction that organisms must not be seen as shadows of ourselves, relatively flimsy or filled out according to their taxonomic nearness, or as objects for our ethical instruction.

In introducing his "Catalogue and description of the natural and artificial rarities belonging to the Royal Society," Nehemiah Grew (1681) addressed himself to the proper arrangement of specimens in museums and organisms in books. Grew began by providing, in our opinion, the finest epigram for a museum collection of organisms—he called it "that so noble an hecatombe," properly acknowledging both the utility and the sacrifice.

Grew then chose a conventional ordering by the *scala naturae,* or chain of being for, as he states, "the very scale of the creatures, is a matter of high speculation." He begins with humans, then descends down the ladder of vertebrates from those with many toes (not because rats are so noble but because monkeys also fill the bill, and consistency demands that one take the bad with the good), to those with hoofs, to those that lay eggs—and then further down from birds, through fishes, shells, insects, and plants. Grew then contrasts his system with those chosen by the two greatest compilers of the preceding, sixteenth century: the Italian Ulysse Aldrovandi and the German Konrad Gesner. He rejects Aldrovandi for too bald an ordering in human terms: "I like not the reason which Aldrovandus gives for his beginning the *History of Quadrupeds* with the horse; *quod praecipuam nobis utilitatem praebeat*" (what exhibits particular use to us). Indeed, Aldrovandi's order is idiosyncratic in the extreme, beginning with utility, then mixing in judgments of nobility and approximation to human form—finally yielding a motley arrangement based on value, beauty, and similarity titrated by some formula known only to himself. His *Ornithologiae* (birds) of 1599, for example, proceeds from the noble (eagles and hawks) to the wise (owls), to the like (bats), then merely to the big (ostriches), the awesome (gryphons), to parrots, crows and, lastly, to small things that go tweet.

As for Gesner, Grew remarks: "Much less should I choose, with Gesner, to go by the alphabet." Here we find a man apparently after our own hearts. For, in the *Historia animalium. Liber 1. De Quadrupedibus viviparis* of 1551, Gesner does proceed right through the latin alphabet: *De Alce* (elks), *Asino* (asses), *Bove* (cattle, *Camelopardali* (giraffes), and so on through *Talpa* (moles), *Tigride* (tigers), *Urso* (bears), and *Vulpe* (foxes). But even Gesner could not avoid the allure of the *scala naturae* and its hidden dictates of relative worth. So, when *Liber II* rolled off the presses in 1554, he treated the cold-blooded quadrupeds again alphabetically (from *De Chamaeleodonte* and *de Crocodilio* to *de Testudinibus,* or turtles), but in a mere 110 pages compared with the 1,104 devoted to warm-blooded creatures. Hence, by relegating them to second place, and much shorter treatment, Gesner disparages animals further down the conventional ladder of life and manages to impose upon his more democratic alphabet the traditional ordering as departure from human worth.

We must confess similar biases. You will find herein more photographs of monkeys than their taxonomic abundance warrants. You will also see few invertebrates, though insects are some 80 percent of described animal species. Aesthetics and instruction may demand sufficient homology of structure to incite sympathy. Still, we delight in *Xenophora,* the bearer of strange things, and in the exquisite shrimp of Solnhofen (see U)—though we do regard as beautiful (see M) the feet of them who preach (we hope) the gospel of peace.

Illuminations

Form and structure do not always coincide. The egg, shaped in its descent through a constraining oviduct, emerges with streamlined symmetry—no edges, no corners to cut soft tissue or snag in motion. But eggshells break with jagged edges, as the physics and crystallography of their structure dictate. The edges are as conspicuous and jarring to natural predators as to us. Niko Tinbergen, our greatest contemporary ethologist, noted that nearly all birds take great pains to remove quickly from the nest all broken pieces of shell after babies hatch. Hawks eat the shells; grebes carry them out of the nest and drown them in their ponds; most birds drop them far away in flight. In a beautiful and elaborate series of experiments with gulls, Tinbergen proved that the thin, broken edge of the shell acts as the triggering stimulus to removal. Thus, the contrast that jars us between sharp break and organic roundness also alerts predators and triggers a parental defense against them. Perception must have a universality transcending culture. The insect that looks like a leaf, a twig, or a piece of dung fools us as well as its natural predators.

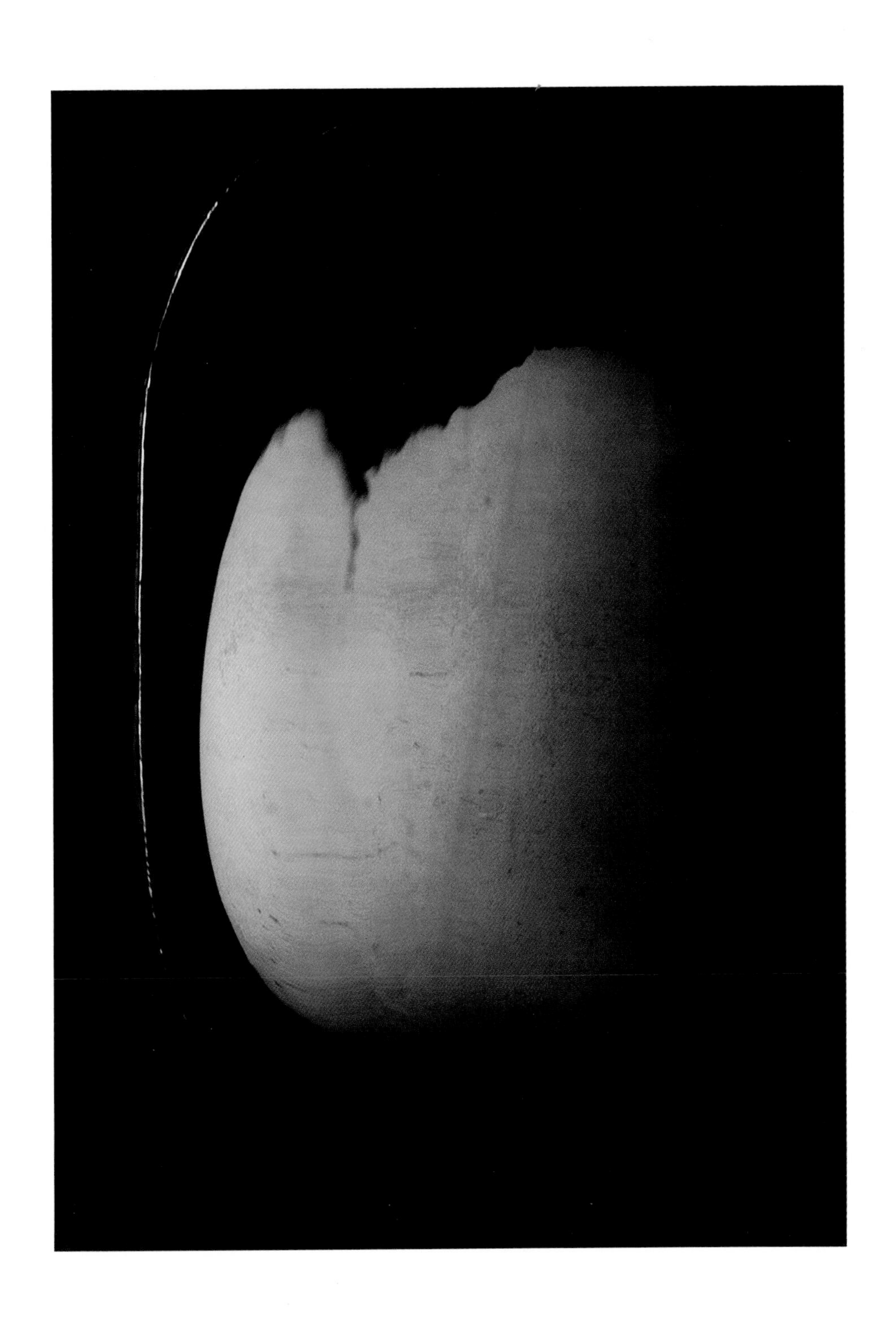

ALBATROSS / *Phoebetria palpebrata*

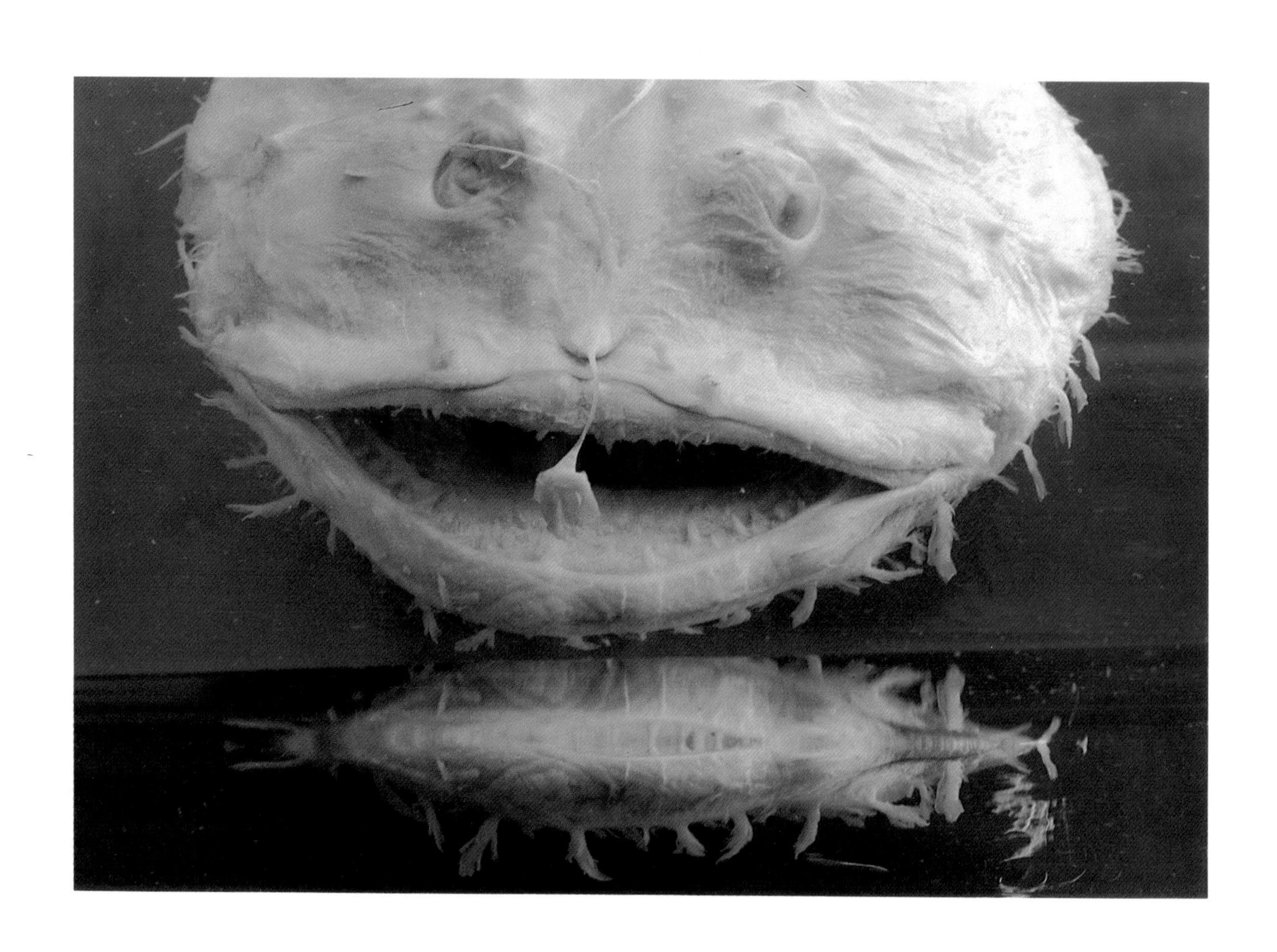

22 ANGLER FISH / *Lophius piscatorius*

An angler fish, bleached white by the same preserving fluid that fills some people to the gills, and reversing its basic plane of symmetry by reflection in its museum jar, sports, limp in the foreground, its defining feature—a "fishing" lure dangled in front of its mouth to attract prey.

Evolution is a quirky tale of old structures pressed and altered to new functions. Angler fishes have detached the first doral fin ray, used in ancestors for support and stabilization of motion, moved it forward over the mouth, and evolved at its tip a set of fishing lures. Some look like worms and glow with the phosphorescence of bacterial symbionts. The lure of one species has evolved to mimic a little fish, complete with eye spots and fins. If I could write science fiction, I would capture the fascination of size and scale by imagining that the pseudodorsal fin of this lure liberated its frontal part to migrate forward and begin a process of recursion.

 BAT / *Rhinopoma microphyllum*

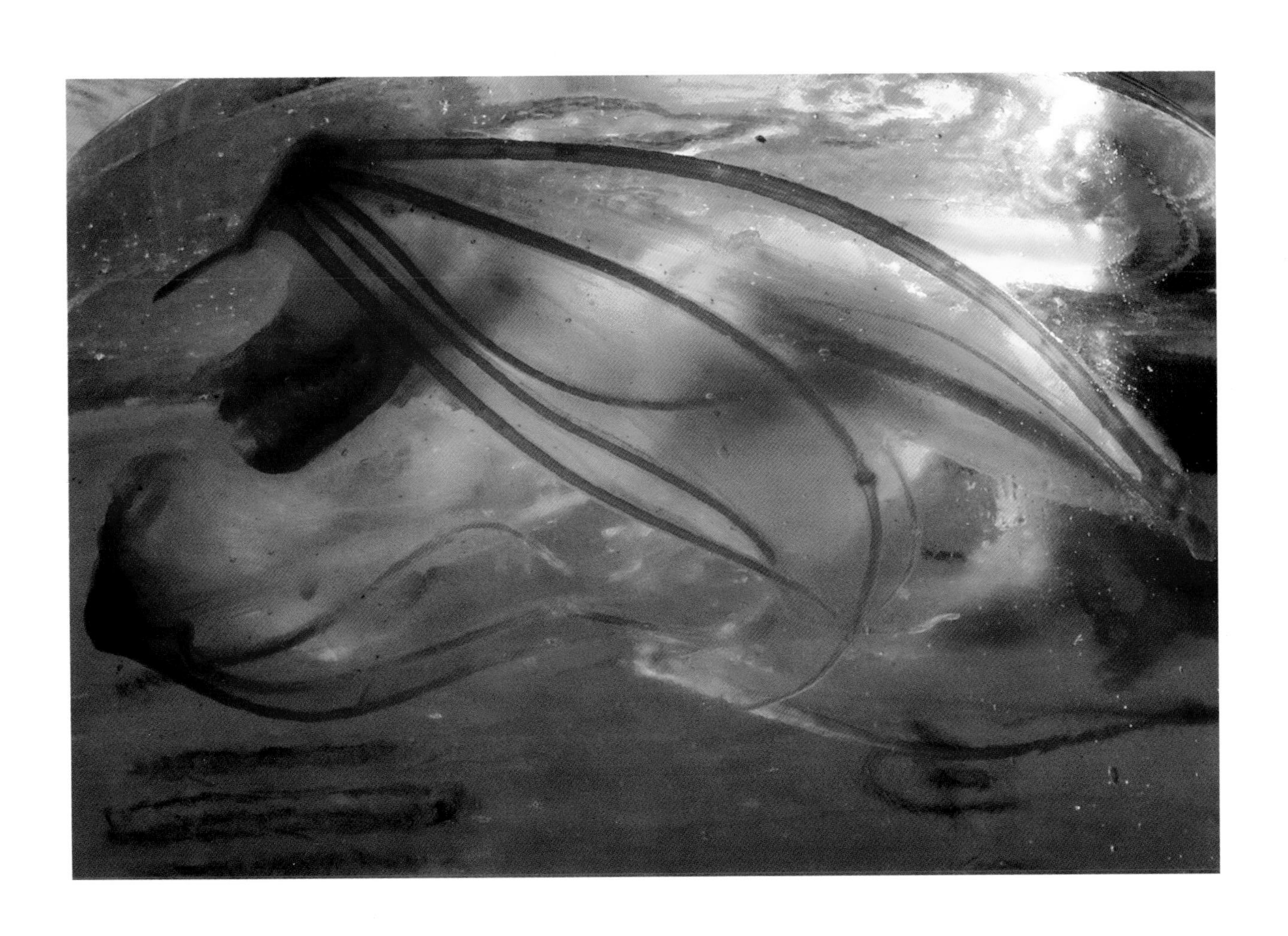

BAT / *Rhinopoma microphyllum*

Like Lugosi drawing his cape to ward off daylight (or the sight of a cross, or a mirror that would reveal his insubstantiality), this bat—unwittingly so constrained in its collection bottle—seems to mock the most famous human legend about its kind. (Most bats, including this pteropodid, are harmless or positively helpful insect eaters of the night. The vampires are a small group of South American bats; they rarely attack humans, though they can injure or even kill unprotected farm animals by the cumulative effect of their individually modest withdrawals of blood.

Our mixed (if inconsistent) perceptions of kinship with and fear of bats may claim an extended pedigree, even within science. In the definitive tenth edition of his *Systema Naturae* (the founding document of animal taxonomy), Linnaeus wrongly included bats in the order Primates. By his description, the order included but four genera: *Homo* (for us), *Simia* (for all monkeys and apes), *Lemur* (for the prosimian primates, lemurs, galagos, and their allies), and *Vespertilio* (for bats)—with only the bats misplaced. Linnaeus's descriptions are usually dry and confined to sizes, shapes, and numbers of anatomical hard parts. Yet, in describing his first species of bat, the vampires naturally, he adds: *noctu haurit sanguinem dormientium*—by night, it drinks up the blood of those who sleep.

Linnaeus, by the way, named but seven species of bats—although bats are second only to rodents in diversity among mammals.

28 BEETLE / *Goliathus*

These Goliath Beetles, giants of the insect world, won their name for obvious reasons. The good humor in this photo might be missed by those who do not know how insects are treated in museum collections. They are pinned through the body onto boards or box bottoms, all (generally) in neat rows, like ranks of soldiers at rigid attention. Slender arrows of steel, thinner and more flexible than the finest sewing needles, are used to transfix the specimens—and these can be bought in quantity from any entomological supply house. But, faced with this behemoth of beetles, Harvard's curators used ordinary nails—and their heads stand out as shining points of metal in an organic matrix.

Some 80 percent of named animal species are insects—yet, despite this unparalleled iteration of opportunity, no insect has ever become much larger than this creature (thank goodness—for we would be no match for the elephantine ants and spiders of so many horror films). The reasons for limitation in size are structural. Consider one argument among many: With a rigid external skeleton, insects can only grow by molting—by casting off the outer shell, permitting the inner soft parts to expand and secreting a new and larger girdle, the source of subsequent restriction (while we can continue to grow around our internal skeletons). A mass of soft parts the size of a rhino could not survive this crucial period of no support—and a giant insect at this critical moment would devolve into a flattened mass of jelly. Lobsters, supported by the sea's buoyancy during this phase of softness, can grow much larger than any terrestrial arthropod—suiting our uses, once again, to nature!

Beer cans are cultural symbols (friend and foe) of natural history for two reasons. They are the remains, first of all, of the primary sustenance of fieldwork. Second, they are the ubiquitous reminders of human intervention even into the most remote and peaceful places of nature's apparent hegemony. But nature fights back effectively and eventually destroys even the "nonbiodegradable." Here, a cover of barnacles begins the process by softening the industrial edge with natural contours and reclaiming the object as an organic substrate.

Most organisms pass even more quickly to oblivion after death. But an occasional bone or shell becomes sequestered in sediment and resists the ravages of time, often for millions or billions of years, as a fossil.

Here we juxtapose (and meld) destruction and preservation by placing atop the beer can three fossil shells (of my own favorite genus *Cerion*) entombed in bits of cemented dune.

CERION / *Cerion agassizi*

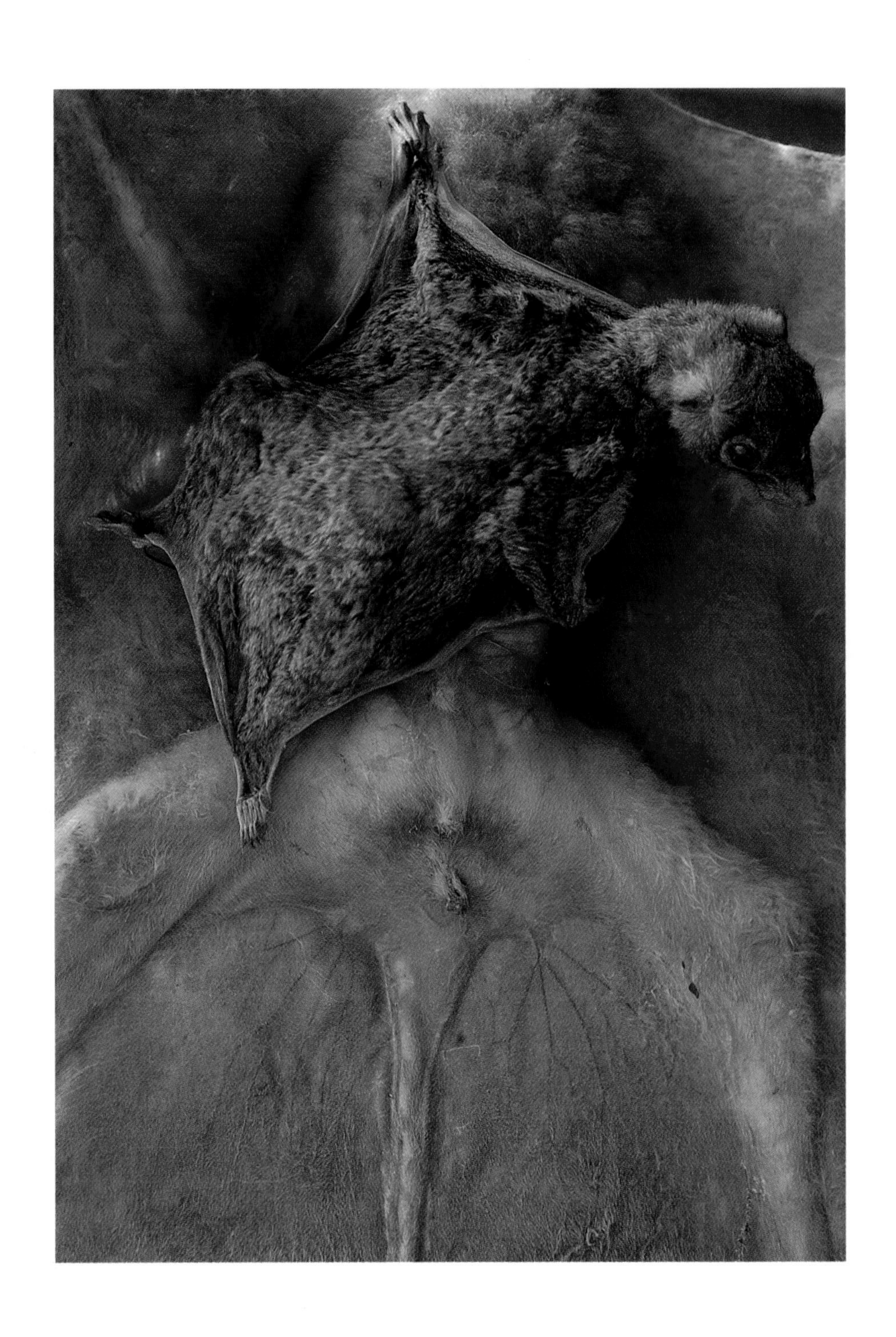

 COLUGO / *Cynocephalus volans*

These so-called "flying lemurs" are neither. They glide and they belong to a small mammalian order all their own, the Dermoptera, containing but two living species, both from southeastern Asia. Gliding may seem an odd adaptation for animals of terrestrial origin, but it has evolved three times independently in mammals (and perhaps a fourth as an intermediate stage in the evolution of true flight in bats)—and it represents an efficient and graceful way to move about in trees. Gliding has also evolved in flying squirrels (rodents) and in a large group of Australian marsupials (gliding possums). But colugos are champs for extent of the gliding membranes. The other two groups stretch their membrane only between the limbs (with fingers, toes, and tails free), while colugos invest all limbs in a continuous membrane stretching from neck to tail—producing a stunning contrast in this photograph of nakedness and efficient covering. Glides may average 100 meters, or more than 200 times the body length.

Eyes decay first after death. The preparator reverses nature and, with human materials, gives greatest stability—and, in this case, centrality—to nature's evanescence.

CAT / *Felis catus*

36 CROCODILE / *Crocodylus acutus*

CHAMELEON / *Chamaeleo dilepis*

38 DOLPHIN / *Delphinus bairdi*

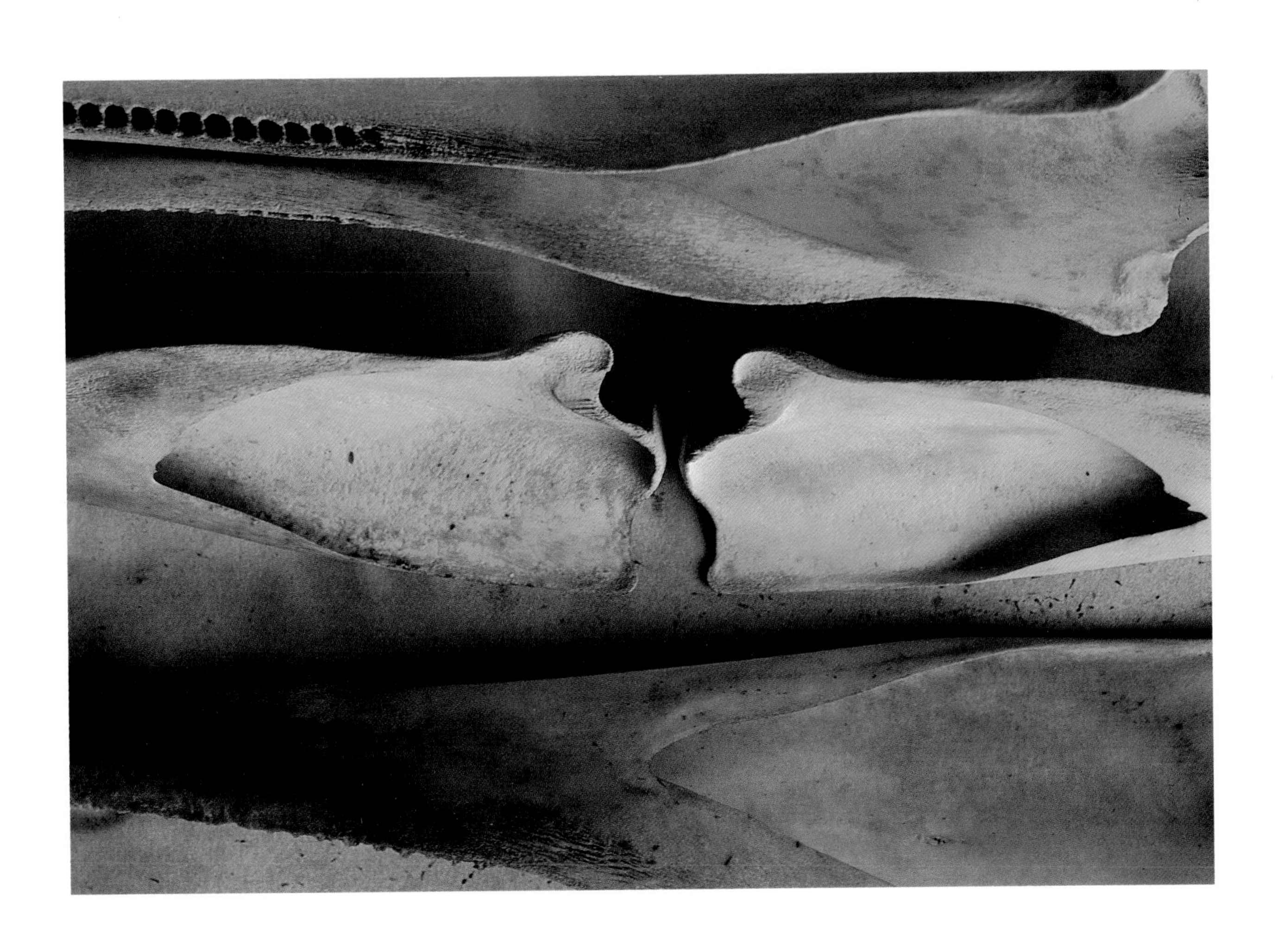

DOLPHIN / *Stenella coeruleoalba*

40 DIMETRODON / *Dimetrodon*

D*imetrodon,* the sail-backed reptile, is included with every kiddie set of plastic dinosaurs. We therefore usually invest this creature with the same label that we apply unfairly to dinosaurs—primitive failure. But *Dimetrodon* is no dinosaur; it belongs to a group that includes the direct ancestors of mammals. Welcome cousin.

Only partly excavated from its matrix (and still bearing remnants of the swaddling cloth that cradled its transition from field to museum), this fossil skull smiles with a sign of relationship. Most reptiles sport a long row of teeth, all much alike. Mammals specialize their dentition to incisors, canines, premolars, and molars. By experimenting with variation in size and spacing of teeth, *Dimetrodon* points the way to its future.

A remarkable and fortuitous similarity of shape links the entire outside, ear against body, with a cross-section through the porous structure of an elephant's skull (where one junction seems to mimic the slope and form of the ear).

In another sense, however, we can specify a deeper, causal link for another aspect of this accidental similarity—and it flows from the most outstanding of all elephantine properties: size. Elephants are the largest terrestrial mammals. Large animals tend to live longer than smaller creatures. This elephant, old by virtue of its size, has experienced all the shocks and minor traumas of a long life, thus

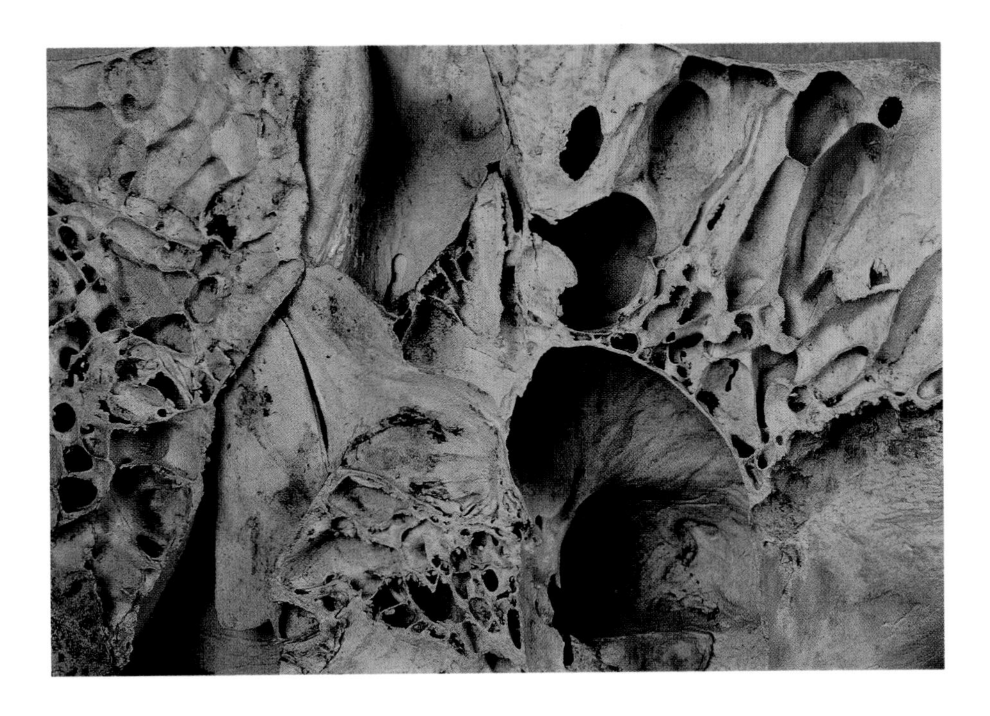

producing the frayed and weather-beaten appearance of skin and ears that sets so much of the similarity we perceive with the complex structure of skull bones. As another consequence of size, elephants live near the limit of supportable weight on land. Life near the edge demands a reduction of weight whenever possible. The massive head (requiring such extensive musculature for support and motion) becomes a primary target for evolutionary lightening—and the spongy texture, producing the fortuitous resemblance to ear and skin, thereby arose as another consequence of size. Age and enlightenment join.

44 EURYPTERUS / *Eurypterus fischeri*

Eurypterus, like the death's-head on an early New England gravestone, peeks forth from its entombing sediment as a revelation from scientific craftsmen early in our century—fossil preparators with dynamite and pickaxes for collection, hammers and rockcrushers for discovery, and tiny chisels and dental picks for detailed exposure.

Eurypterids, denizens of fresh water lakes, are the largest arthropods that ever lived. This head shield seems to show eyes, a clear nose with side lobes, and a grim mouth. The eyes are correct, the rest illusion in human terms. We see the top of a eurypterid head shield. The eyes look forward (upward in the photo), not at us. The "nose" is a bump on the midline of the upper surface, the "mouth" a junction between head and body plates of a segmented animal. (The true mouth lies underneath, still hugging the sediment, and unliberated by the preparator.)

A famous Japanese crab seems to have a samurai face engraved upon its carapace. Some overenthusiastic devotees of good order in nature, forgetting the fundamental truth of history's quirkiness, have attributed the face to an odd manifestation of natural selection—as fishermen return to the water those specimens accidentally varying in most facelike directions. But arthropods, including crabs and eurypterids, are bilaterally symmetrical animals, with various bumps, dots, and creases (for their own reasons) along the midline and in symmetrical patterns on either side. Human faces are also the product of bilateral symmetry under such constraints. Purely accidental resemblances are bound to arise from time to time. Fishermen might, at the very most (given the tiny fragment of geological time that we occupy), slightly enhance a resemblance already set for no direct reason. The earbones of a certain whale resemble, every bit as well, a Norwegian fisherman smoking his pipe—but no one ever returned Leviathan or Moby Dick to the waters because its invisible malleus inspired reverence.

The first bony fishes evolved a rigid skull with upper jaw firmly fused to cranium. Such fishes must surround their prey with their mouths. As an outstanding trend in the evolution of most modern fishes, the upper jaw becomes highly mobile. The most forward bone, the premaxilla, bears most or all the teeth. In most vertebrates, the maxilla just behind serves as the large and primary tooth-bearing bone, but in these advanced fishes, it becomes a rod loosely attached to the premaxillary jaw. The premaxilla also loses its rigid connection to the cranium above, replacing the old bony fusion with a flexible ligament. Thus, the upper jaw (premaxilla) has only flexible connections to the maxilla below and behind, and to the skull above. These advanced fishes need no longer envelop their food with mouth and body. Instead, they throw their jaws at their food, as the maxillary bar pushes the toothed premaxilla forward—a substantial advantage for capturing prey with maximal speed and minimal danger. This so-called "protrusibility" benefits the fish for reasons just cited, but it also sets the humor and fascination of these photographs. The numerous mobile elements and the wide gape that they permit leads us to view these fishes as engaged in earnest conversation. Again we are fooled by our drive to read nature in human terms—though protrusibility does speak to us in other ways.

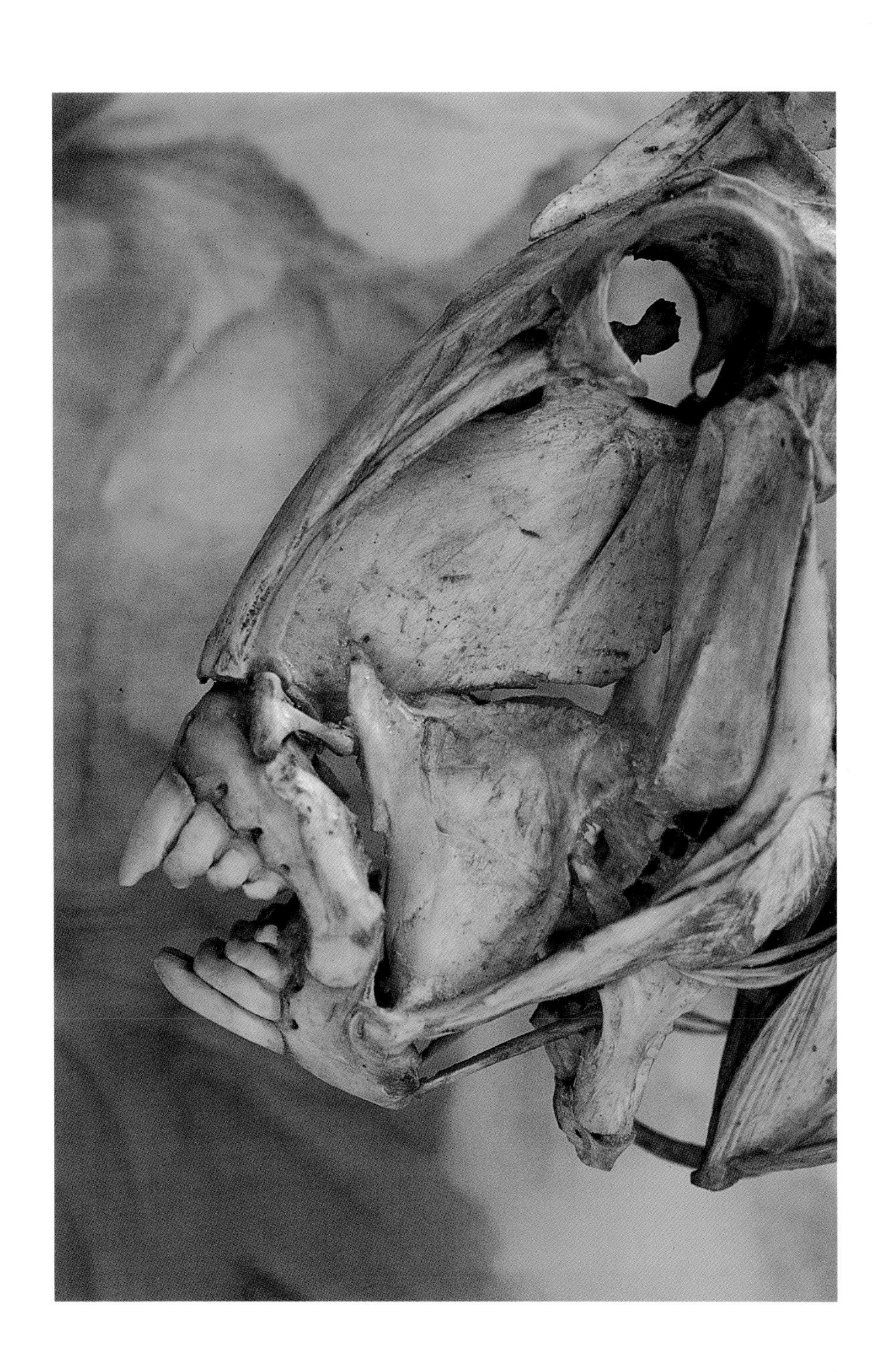

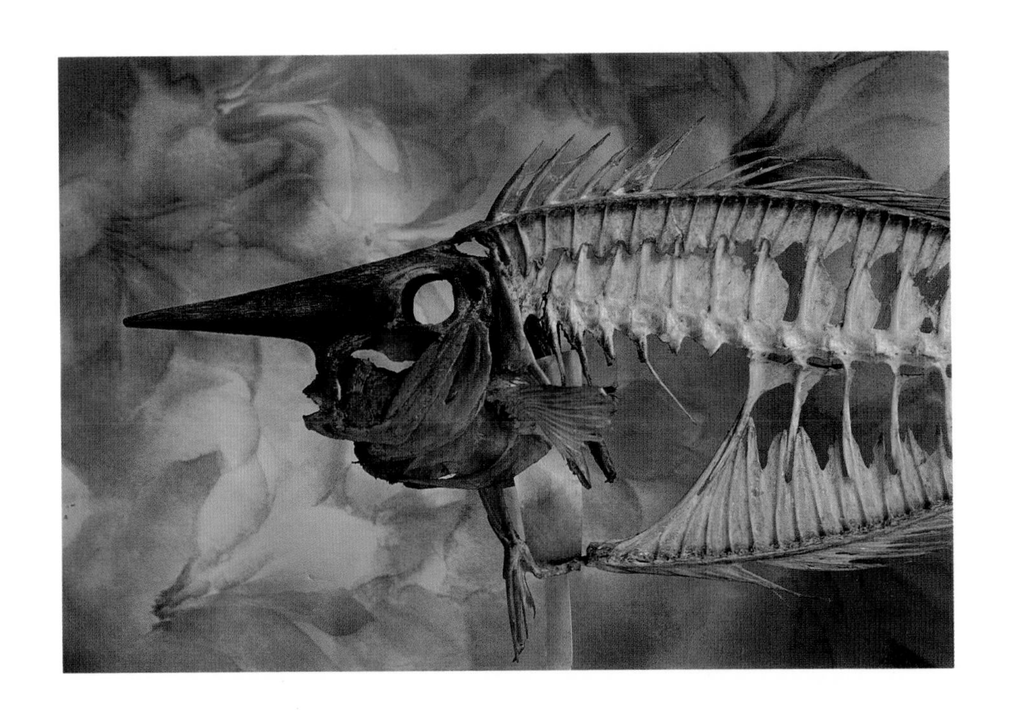

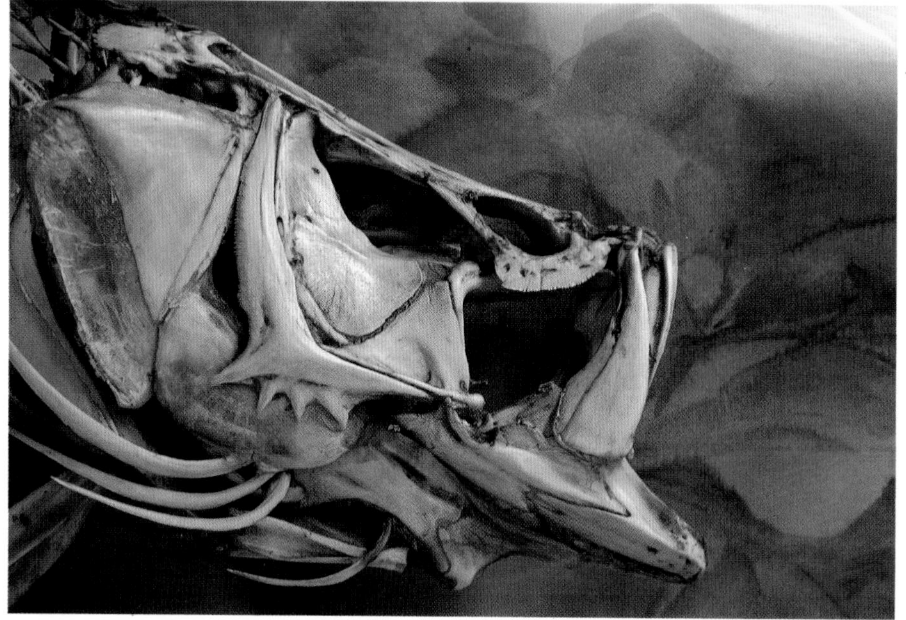

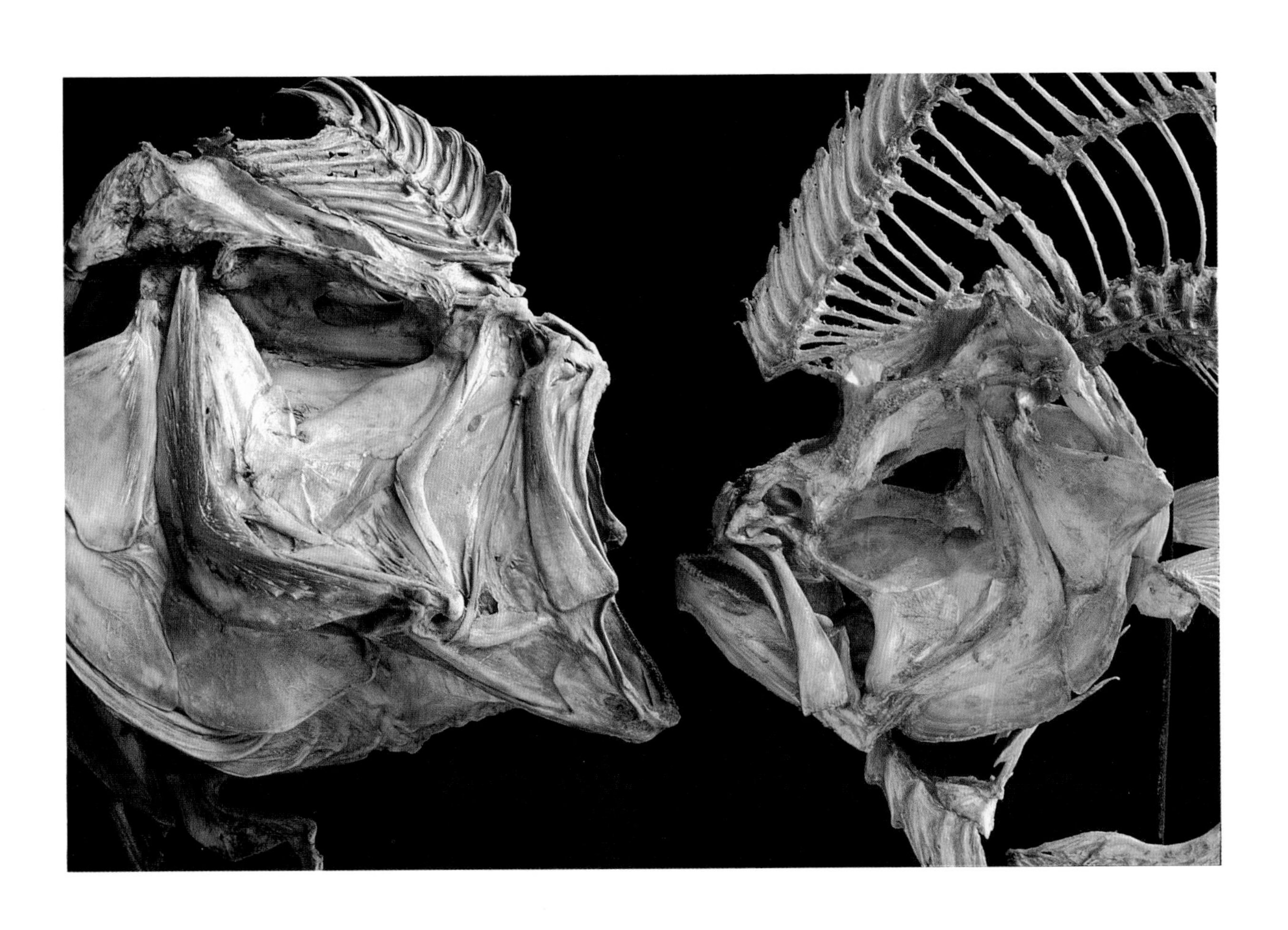

Flamingos are named for their flaming red color (flamenco dancing has the same root). But colors fade after death and only become bright in the first place when flamingos can find food with properly convertible chemicals—thus proving the old motto, *Der Mann ist was er isst* (you are what you eat). That most preciously absurd of all biological notions—that flamingos evolved their red color to gain protection from enemies by fading into invisibility before the sunset—fails from this observation of inconstant redness, among other reasons!

Flamingos feed by swinging their heads down between their legs and filtering small creatures from the water on an elaborate mesh of ridges and grooves evolved within their bills (and using their thickened tongue to pump water in and out). Flamingos feed in reversed position, head upside down, upper bill on the bottom. This unusual orientation has engendered a complex evolutionary rearrangement. The flamingo's upper bill now looks like a lower jaw—long and slender, as it slots into the massive lower bill serving as a stout buttress (like the upper bill of most birds). Flamingos also feed by moving the upper bill up and down against the stable lower jaw. The flamingo's smile (of the upside-down head) records its capacity to fool us—for we readily read the shapes of the bills as we know them on ordinary birds.

FLAMINGO / *Phoenicopterus ruber*

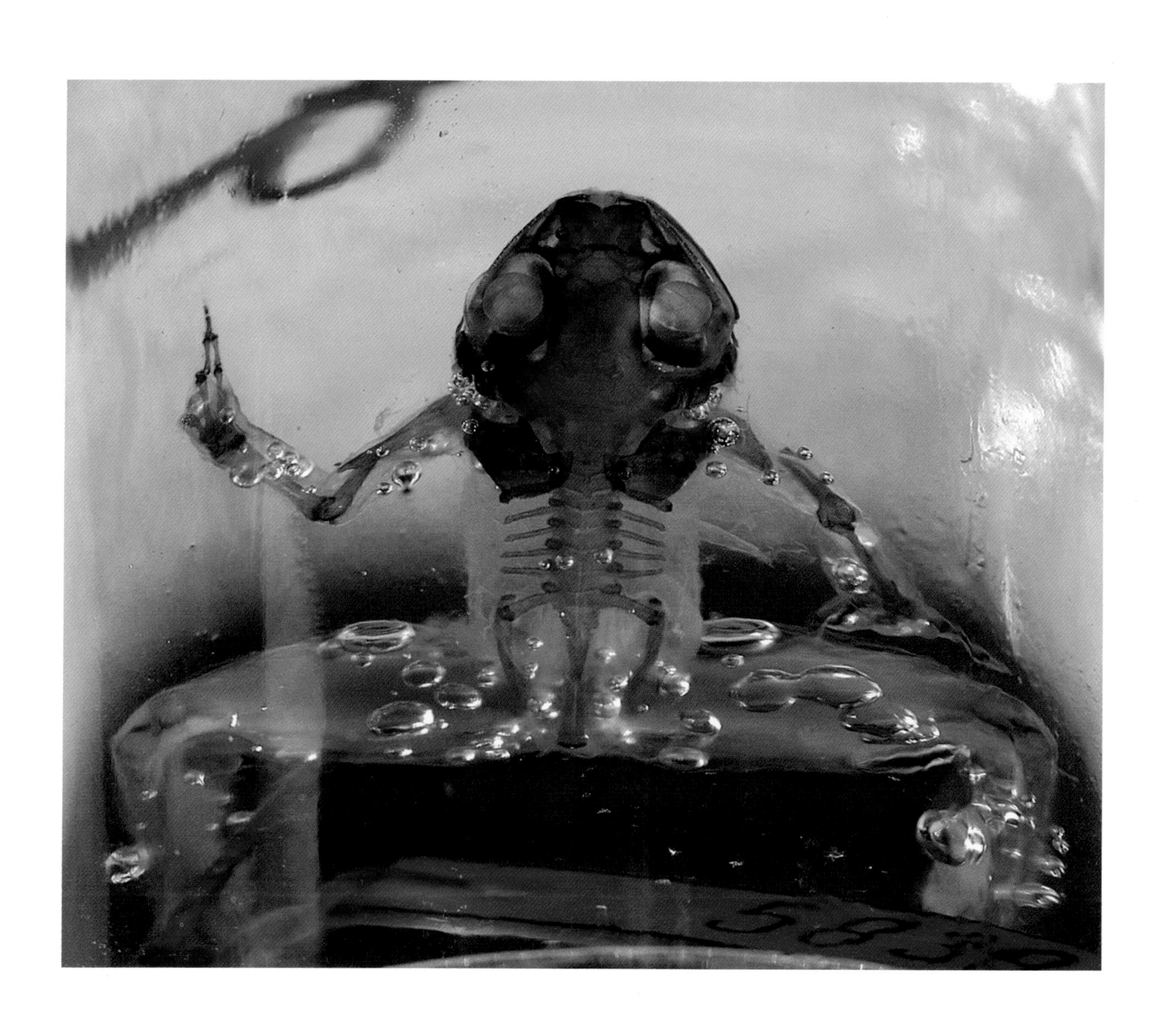

54 FROG / *Colosthethus bocagei*

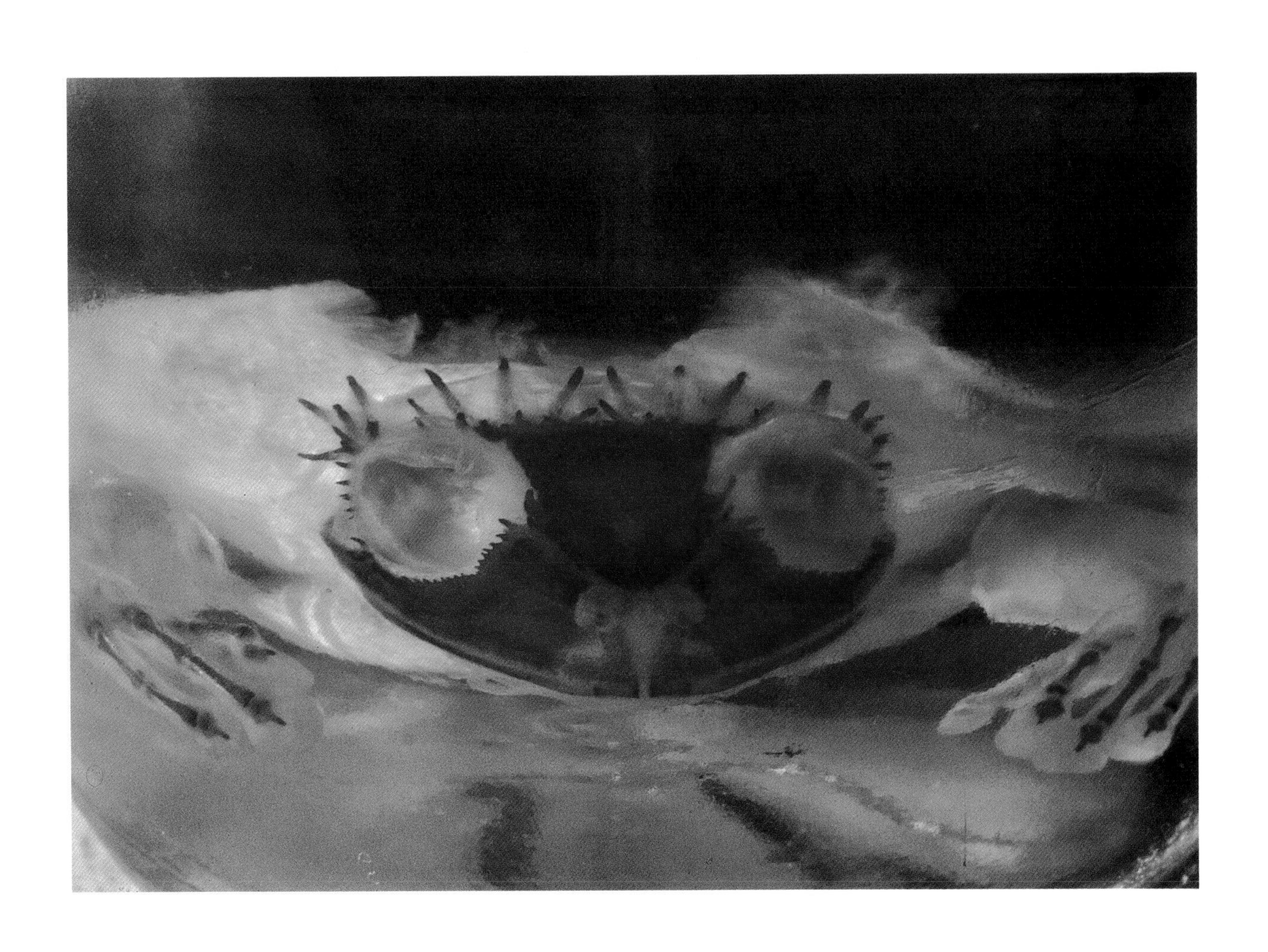

FROG / *Anotheca spinosa* 55

Gannets belong to the family Sulidae, a group of sea birds including the boobies and famed for a consequence of behavior that we regard as stupid. Sailors gave boobies their name because the birds would fly close to ships and make no attempt to escape club-wielding seamen. (Such birdbrained activity saved many a naval party in distress, including Captain Bligh and his loyalists in the long boat during their forced 4,000-mile journey across uncharted seas.)

We make intuitive judgments about "intelligence" in animals (often most inappropriately) by their degree of approximation to the two great differences that human heads have evolved: a rounded and high-vaulted cranium and a short face. With their long beaks and flattened crania, sulid skulls seem to deserve their name and reputation. This photo, however, appears to mock such a self-congratulatory attitude. We see the ghostly double reflection of a gannet's skull on the glass surface of a bell jar placed above it. The curvature causes one reflection to assume those very features—short face and round cranium—that we deem a sign of intelligence. Is this a double inversion of the metaphor of Plato's cave? Is the real skull the shadow and that humanoid reflection the archetype? Or is the shadow only the reflection of our deepest cultural bias?

56 GANNETT / *Morus bassana*

GANNETT / *Morus bassana*

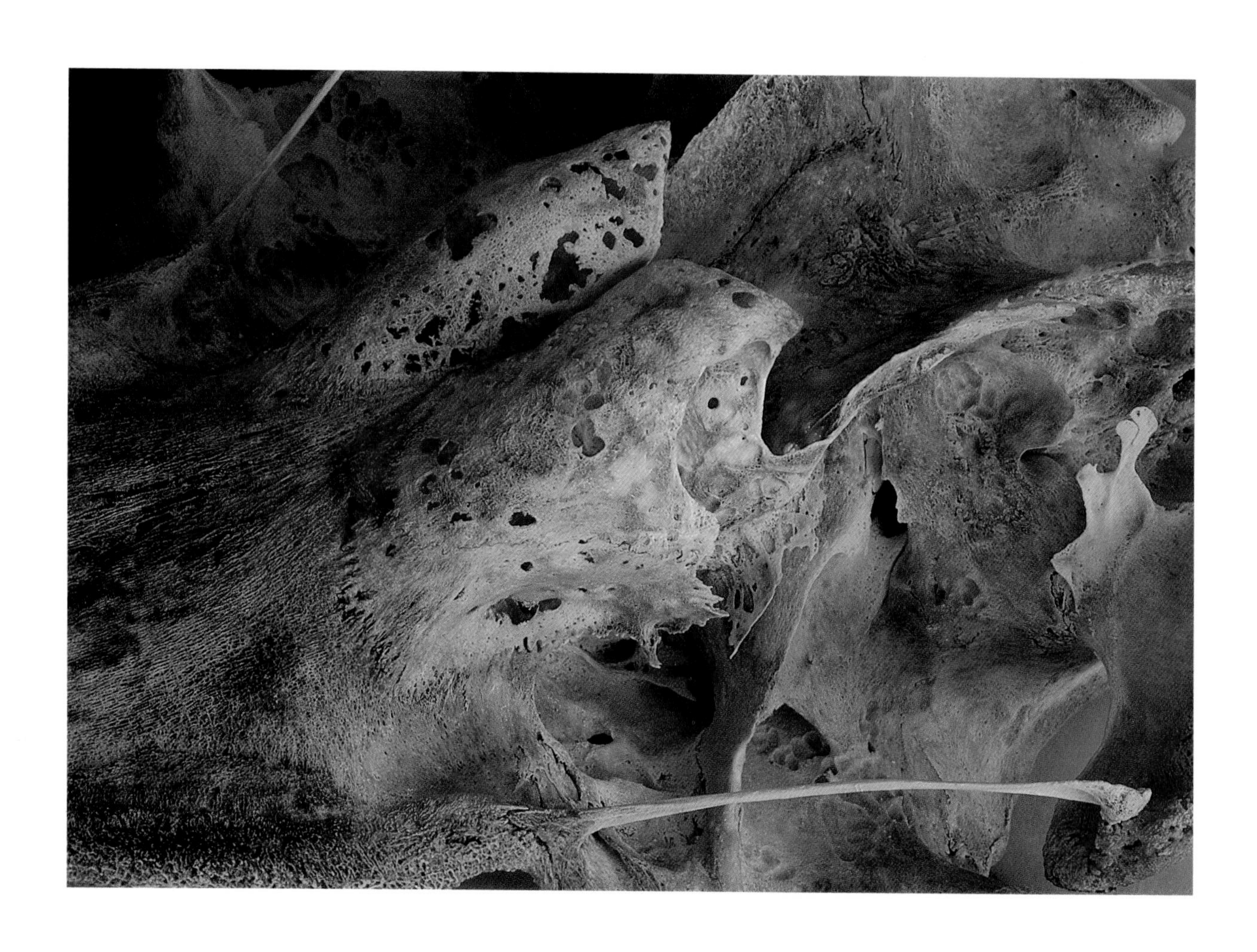

58 GRAMPUS / *Grampus griseus*

HARE / *Lepus americanus*

"Gentille alouette" is a lark, but *Alouatta,* the South American howler monkey, scarcely merits the near homonymy. Numerical comparisons of *Alouatta* with other new world monkeys have identified this genus as a "phyletic giant"—an animal that has grown beyond the usual trajectory for its group, thereby accentuating features that are increasing, but stop sooner, in related species. This status explains both the massive lower jaws and the small brain relative to body size. (Jaws increase more rapidly, brains more slowly, than bodies as primates grow. *Alouatta* has simply extended these common trends.) This massive jaw also underlies the double take of this photograph.

We seem to see a face looking at us, but it is only pseudo-Janus—for we are looking through the rear end of the skull (the circle on top is the giveaway, for we do not witness a trepanation of the skull, but the foramen magnum or "great hole" that articulates skull with spinal column). The mouth and teeth are the animal's own (though we see them through the skull, and in the distance), but the corners of its "hat" are the arches of its upper jaw, while the hunched "shoulders" are the massive lower jaw.

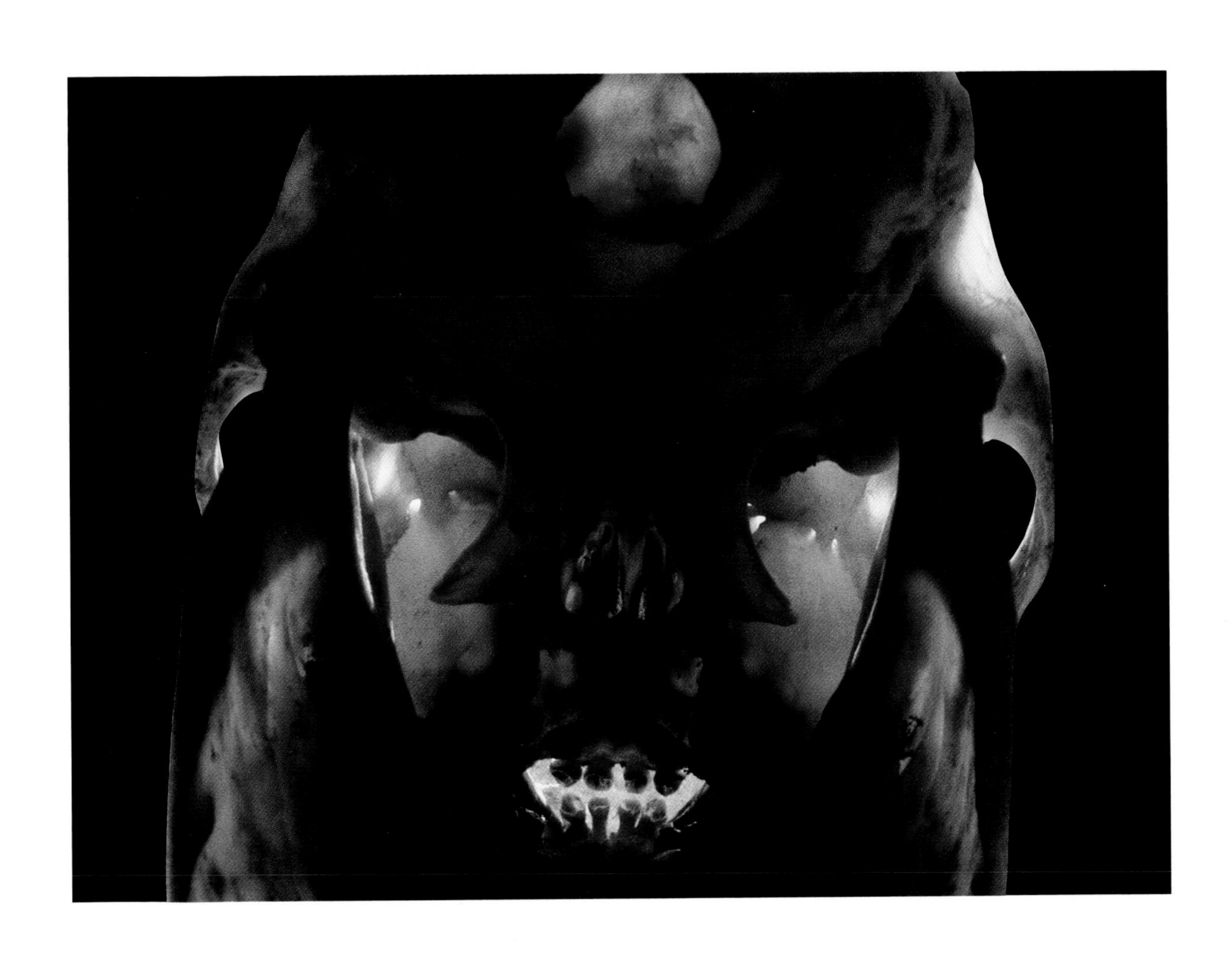

HOWLER MONKEY / *Alouatta belzebul*

In a rare, direct imposition upon modes and realities of preservation in museums, Rosamond has playfully drawn circles in the universal patina of all collections—dust. Thus, the banded contrast of nature and storage combines with the natural form of eggs to force these disappointed containers of future ibises into similarity with the beach-rounded pebbles made of dark igneous rock intruded by bands of white quartz or calcite—and so often found on New England's beaches. Eggs achieve their streamlining by direct shaping in the oviduct, pebbles by the opposite route of erosive sculpturing from original roughness—but the common form does reflect a higher similarity of forces.

Since dust pervades museums, and seems both ineradicable and constant, sleuths may use its thickness as a guide to the age and stratigraphy of collections. I once opened a drawer in an old part of our collections. The contents had been dumped and sheepishly piled back in disarray—but obviously very long ago as the dust testified. I found a note, also encrusted by the universal patina. It was dated 1861 and contained both an apology and exculpating explanation, penned by a terrified student lest the intense and temperamental boss of the museum, Louis Agassiz himself, discover such a calamity without appropriate documentation. Agassiz, obviously, never opened the drawer. The student, Nathaniel Southgate Shaler, went on to become a famous scientist. The dust still rains and reigns.

IBIS / *Plegadis falcinellus*

Taxonomists have described more than a million species of organisms. The vast majority are known to few, if any, specialists. A handful have distinguished themselves by their centrality to our lives. The placid, but very dead, look of these Indian jungle fowl may not be inappropriate in human terms since they are the direct ancestors of all domestic chickens. We have recruited them for our purposes in ways beyond their obvious utility at table. The wild jungle cock crows at sunset, and no one knows how we reset their descendants as alarm clocks.

JUNGLE FOWL / *Gallus gallus*

The kiwi (genus *Apteryx,* or "wingless") belongs to the small groups of large-bodied flightless birds (ratites) that inhabit all southern continents, but no northern lands (ostriches in Africa; rheas in South America; emus and cassowaries in Australia; kiwis and the all-time giants, the extinct moas, in New Zealand).

Their geographic distribution belongs to the large class of data in the evolutionary sciences that make no sense in terms of current environments, but stand simply as a record of history—in this case, the former connection of all southern continents as Gondwanaland.

The association of large body size with loss of flight, on the other hand, belongs to that other great class of direct adaptations—causal correlations of form with *current* function. Galileo himself first established the basis of this association in his great last work, written by stealth under house arrest by the Inquisition. As objects increase in size, their surfaces (length × length, or $length^2$) must grow more slowly than volumes (length × length × length, or $length^3$). Since flight depends upon wing surfaces that must hold a volume aloft, flight becomes impossible above a restricted size. Albatrosses, the heaviest fliers, may look elegant on the wing, but they can barely maintain themselves by flapping, and "cheat" instead by searching out thermal updrafts to keep aloft. Loss of flight permits expanded size; increase in size demands loss of flight. This association evolves again and again in unrelated lineages, a sign of its functional necessity—the extinct dodo of Mauritius (a giant pigeon), the extinct great auk, and the thriving ratites. The down feathers of the young, and the dust-mop of its parent—the textural basis of "unbirdness" in this photograph—record a further freedom to change once the rigid demands of flight are rescinded.

KIWI / *Apteryx mantelli* 67

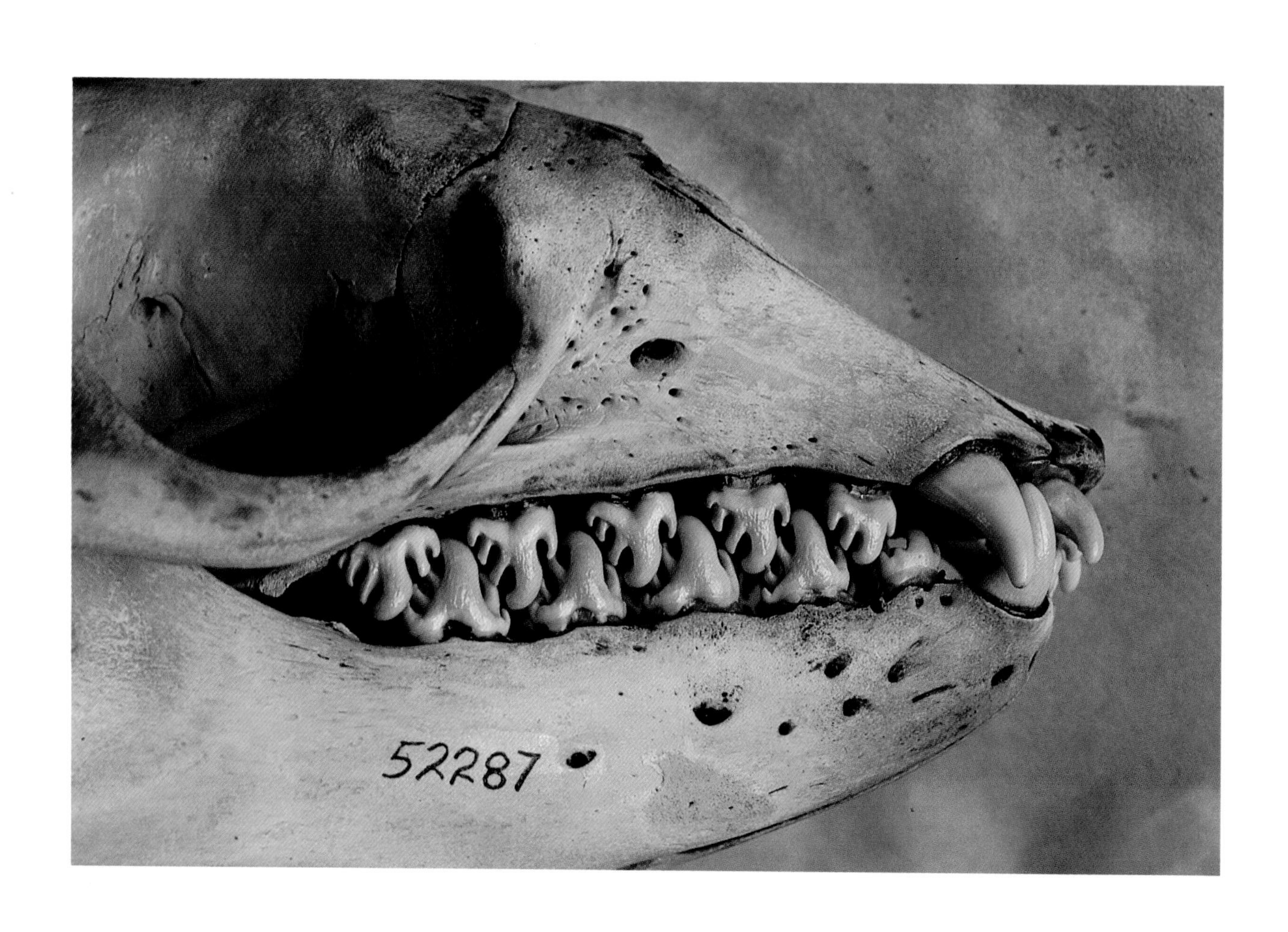

LOBODON / *Lobodon carcinophagus*

Lobodon, literally the "bumpy-toothed" seal, uses the straining mesh of corrugated teeth to filter krill (small arthropods) from the plankton. Modern baleen whales, the largest animals that have ever lived (no dinosaur ever came close to a blue whale in the heavyweight derby), subsist, paradoxically, on the same tiny krill—also filtering them from the plankton, but with huge plates of whalebone. Flamingos, while wading in shallow hypersaline ponds, swing their heads down and, with beaks reversed, also filter small arthropods—on complex ridges and furrows evolved from their bills. Three creatures of no close relationship all make a living in the same way with structures of similar function fashioned from different bits of anatomy—tooth, jawbones, and bills. Such convergence is nature's finest illustration of adaptation—good fit between form and function, built differently because evolutionary pasts must peek through current excellence.

Rosamond and I agreed at the outset that we would trust each other's different professionalisms and would not question choice of photos or textual themes. Yet we argued more about this photo than any other. I found the identification number ordinary and discordant—the ubiquitous method used by museums to catalogue specimens and, incidentally, so often to destroy their aesthetic integrity. She found it striking and unusual—a kind of prison signature with many layers of meaning. It is well that an artist and a natural historian should see the meaning of a simple alteration so differently—and as a result of so many years spent thinking in a certain unchallenged way. The only overarching theme of a bestiary is diversity.

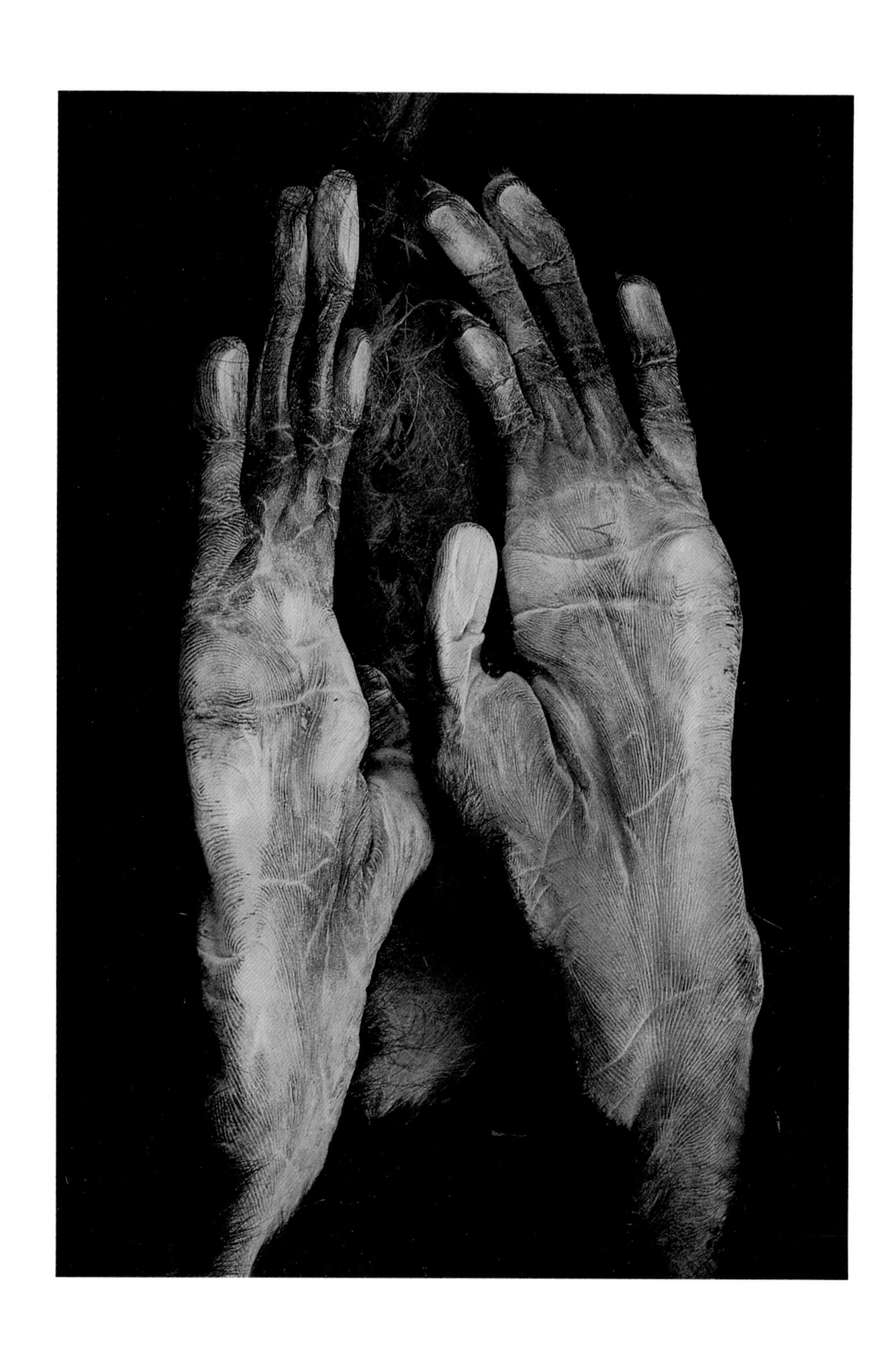

 LANGUR / *Presbytis chrysomeles*

The *rear* feet of a langur monkey and the corresponding parts of an exquisitely preserved nearly 2,000-year-old "bog person" from Denmark remind us that the most fundamental changes in human evolution must be viewed from both ends of the body. The "big two" of our origin are upright posture and enlargement of the brain—but the bottom led (and may have inspired) the top. The earliest human fossils walked fully upright and carried brains little larger than an ape's. Freeing of the hands and subsequent development of tools may have entrained the social changes that made a large brain so valuable.

Nothing so stuns my mind as an image misinterpreted in scale by orders of magnitude because it has no sure reference point in human bodies or artifacts. The Grand Teton Mountains are named for their whimsical resemblance to the female breast. Fossil mastodons (extinct elephants) evolved molars with cusps in pinnacled rows—so another *man* of science named them "breast tooth." This ancient tooth, in the cradle of cotton wool provided by museum curators for its protection, might pass for Wyoming. Size, Julian Huxley once remarked, has a fascination of its own.

Superior beings when of late they saw
A mortal man unfold all nature's law
Admired such wisdom in an earthly shape
And showed a Newton as we show an ape.

Pope wrote these heroic couplets to express how the gifted inhabitants of Jupiter (a heavy planet sure to evolve weightier intellects) might display the paragon of earthly achievement. The film *2001* ends with a different version of transcendence directed from Jupiter. The human hero, having tasted higher things, floats down as an embryo in its amniotic sac, to the tune of *Also Sprach Zarathustra,* Nietzsche's dubious paean to improvement. Do we treat our nearest primate relatives as Jupiter, in literature, twice managed us?

NIGHT MONKEY / *Aotus trivirgatus*

 OWL / *Otus asio*

The usual conjunction of owls and book is the *ex libris* bookplate showing either the head with its large, rounded eyes, or the whole creature sitting serenely on a perch, exuding the wisdom of Minerva.

I find it so consistent with the major theme of this book—new perspectives added by art to information lost in decay and storage—that we see here the skeleton of an owl's nether end perched upon a book retrieved from a junkyard!

To the question "why can't penguins fly?" the proper answer is: "they do—but under water."

Most swimming birds paddle with webbed feet, but penguins flap their wings, as a bird in ordinary flight, to move through water (penguins use their own webbed feet as a rudder for steering, not for propulsion).

Penguin wings retain the heavy musculature of ordinary flying birds, and their breastbones have high projecting keels for attachment of the flight muscles. (Both these features are lost in truly flightless ostriches and their kin.)

Penguin wings are well designed for underwater flight in many ways. The front or leading edge is rounded, the back or following edge (here surrounding the baby) tapered off, thus reducing drag and increasing propulsion. Feathers are reduced to "small, scale-like things" (in the words of G. G. Simpson) that further reduce drag without losing their important role in insulation.

This sharpness of wing edge and ordered arrangement of the reduced feathers give this adult penguin its almost metallic or "industrial" aspect. But the soft baby, not yet ready for the rigors of life as a flying machine in a dense medium, retains a more "organic" look.

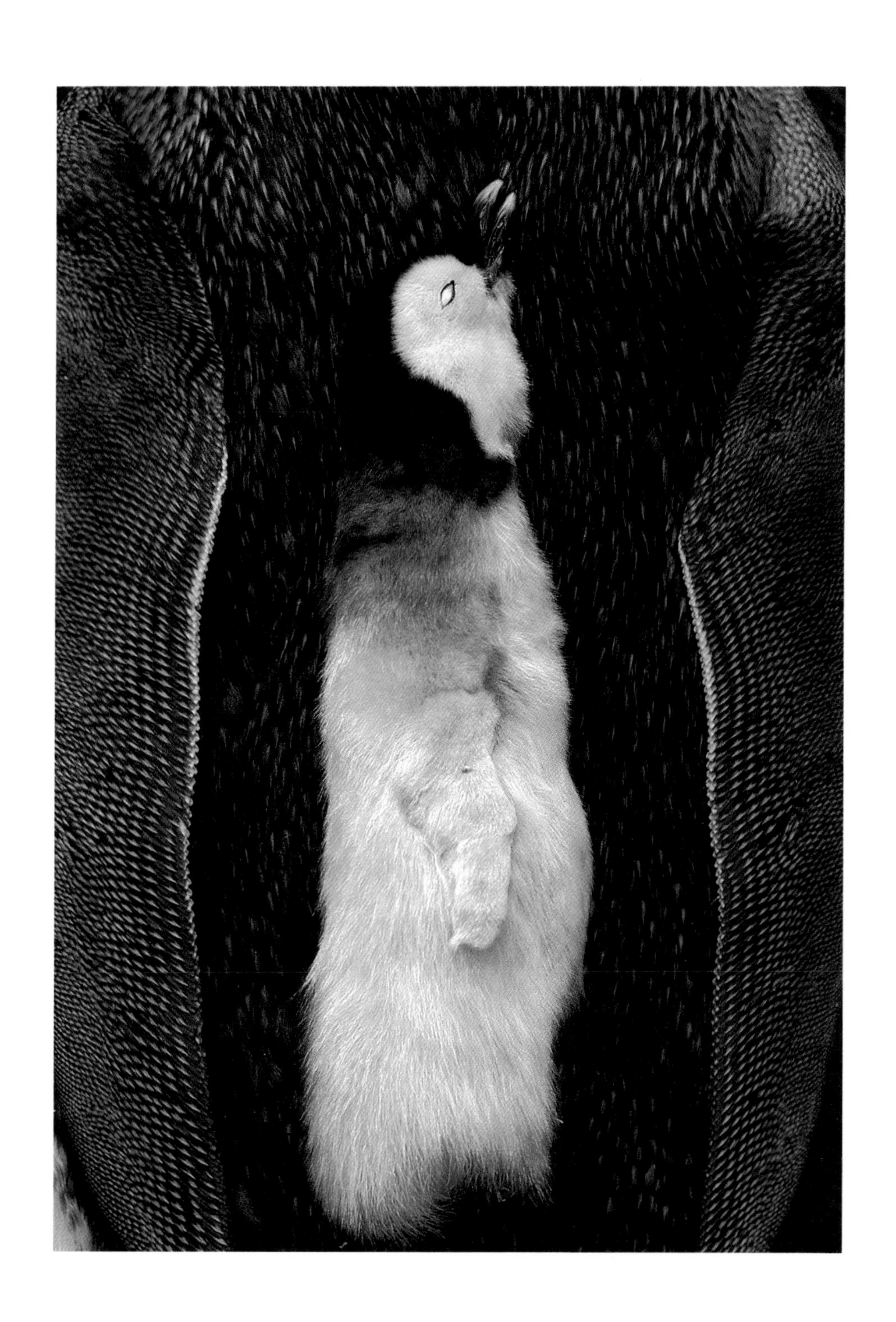

PENGUIN / *Aptenodytes forsteri*

Nasalis, the proboscis monkey, begins life as a snub-nosed youngster and apes Durante as he grows. The relationships of humans with other primates are evident more in the similarities of all primate babies than in the differences that accumulate through growth. Young apes have a rounded cranium and small jaws, but brains grow more slowly than bodies, jaws more quickly—and the gracile baby chimp becomes the low-vaulted, jaw-jutting adult. Humans follow the same pathway through growth, but ever so much more slowly—so that, as adults, we retain important features of baby ancestors—an evolutionary process called neoteny, or retaining youth. We are, in many ways, permanent children. We live longer than any other mammal, mature more slowly, and retain the flexibility and playfulness of childhood. Except ye become as little children, a wise man once said, ye shall not enter into the kingdom of heaven.

Quetzalcoatl, the feathered serpent god of the Aztecs, sported a headdress of quetzal plumes. Quetzals, brilliantly colored birds of the trogon family, live in high forests from Mexico to Panama. Mayans and other native peoples revered the bird, and treated it as a renewable resource. They would pluck the feathers, and return the birds to the woods to grow another set, perhaps for another shearing. European conquerors almost eliminated the species by using outright murder to collect feathers. Quetzals continue to inspire awe, and Guatemala has granted this species a kind of ultimate tribute in our commercial society by naming its monetary unit the quetzal.

The brilliant color of the plumage stands in stark (or should I say bright) contrast with most others in this book, and seems almost to contradict the general theme that death and procedures of collection and storage strip away information—for art to reanimate. The reason is structural and interesting, not a result of recent collection or unusually careful storage. Most animal colors are pigments, chemical substances that will degrade and fade after death. The red breast of the quetzal is a pigment and will soon lose its richness. But the shining metallic green will remain and withstand time's ravages, for the green is no pigment at all, but an iridescence. Its basis is structural, not chemical. Microscopic granules in the feathers break white light into brilliant greens, blues, and golds that shift and sparkle with intensity and angle of illumination. By not being there, in a sense, the green will remain.

QUETZAL / *Pharomachrus mocinno*

Taxonomists love the molar teeth of mammals—for each group displays such a distinctive pattern that a single tooth identifies the animal. (The few ambiguities, however, sometimes produce great embarrassment. Pig and human molars, for example, can be strikingly similar—and *Hesperopithecus,* the ape-man destined to put America on the map of paleoanthropology, turned out to be a swine.) Rhino molars, with their connecting bar between two parallel ridges, like a Greek letter pi, are among the most diagnostic and unforgettable of teeth.

A popular legend grants paleontologists such arcane anatomical wisdom that they can unerringly reconstruct the entire animal from a single bone. We do this, the myth runs, because laws of structure make each piece imply all others in a cascade of logical inference—foot bone connected to the ankle bone, etc. Nonsense. We perform this feat, when it can be done at all, by simple observation and experience—by induction, not deduction. Teeth of this form do not imply horns and thick skin by laws of mechanics. We know that a rhino once surrounded a pi-shaped tooth simply because this distinctive molar has never been found in any other kind of animal but a rhino.

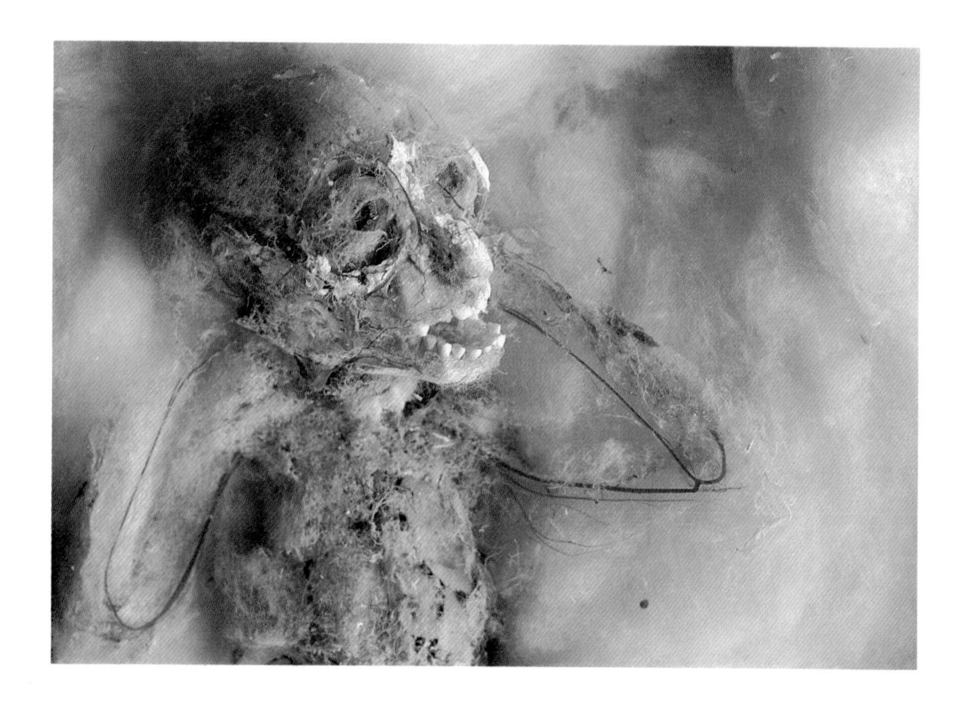

The vertebrate body is an extraordinary jumble of overlapping systems somehow coordinated as a marvelous machine. When viewed all at once, the whole structure is so complex, so "busy" that it defies our understanding. We have to take it apart, system by system—and the pedagogical devices for such analysis are legion. Consider the "transparent" man or woman of our science museums, and the standard mode of public demonstration—bones first, then nerves, blood vessels, lymphatics, organs, muscles, and skin, added step by step from inside out, until we clothe the terror of Halloween with the flesh of aesthetic comfort.

Scientists have struggled for centuries to disentangle the complexity by highlighting single systems and removing all others. Underlying skeletons are easy to retain, overlying soft anatomy more difficult. Many of the most famous "prepara-

tions" of our anatomical museums present just nerves, just blood vessels, or just the lymphatic system. The search for new methods goes on. Here we see two squirrel monkeys treated to preserve the vascular system alone. Blood vessels are injected with a liquid treated by a catalyst that will cause it to polymerize (harden) at room temperature. Surrounding tissues are then dissolved away, leaving a complete image of the animal expressed in but one of its "layers."

Michelangelo insisted that his figures lay within untreated stones and that his job as a sculptor was merely to reveal. These vascular monkeys, photographed against the cotton that cradles them in their actual repository, seem also to emerge from an inchoate background—a visual metaphor for the important truth of evolution as differentiation.

We often err in using the absolute ticking of Newtonian time as a universal clock for all organisms. This imposition of human technology masks important similarities that only become apparent when we interpret the physiology and motion of each animal by its own internal clock. All mammals live about the same time by their physiological clocks—as a mouse in its few months breathes and beats its heart as many times as an elephant during its many decades. Who can impute an absolute value to the fabled lethargy of the sloth (a name imposed upon this South American mammal, discovered long after the seven deadly sins received their labels)? Do they view our normal activities as an inverted movie of the Keystone Cops?

The remarkable flexibility of a sloth's head (so oddly discordant with the limited repertoire of its body)—and a source of the peculiar posture that evokes such a human response in this specimen—arises from its greatest anatomical peculiarity. Virtually all mammals have seven cervical (neck) vertebrae (they are very long, but not more numerous, in giraffes). Sloths have nine.

The Isthmus of Panama rose within the last 2 or 3 million years. Before then, South America was an island, a super-Australia with its own fauna. Most of these oddballs died as North American competitors swept south in waves of migration. But one order thrived as a memory of previous richness—the Edentata, or toothless ones, including armadillos, anteaters, and sloths.

SLOTH / *Bradypus pallidus*

90 SAKI / *Pithecia pithecia*

Saki / *Pithecia pithecia*

The so-called "eye spots" on the backs and wings of many insects may serve several possible functions. In some species they loom large, as if truly the eyes of a much bigger creature—and may thereby scare predators away. In others, they lie at an expendable periphery, attracting a predator's attention to wing tip and the air beyond—and not to the vulnerable body. In any case, the enormous variety of circles, bars, and splotches invites our own playful interpretation, whatever the function (if any) for the insects that carry them. One naturalist found in moth and butterfly wings all the letters to reproduce a line of Roethke's poetry—"All finite things reveal infinitude." Here, arrayed like cartoon early worms poking their heads above the ground, we see dark spots at the wing tips of *Samia* moths facing each other in two rows.

SAMIA MOTH / *Samia cynthia*

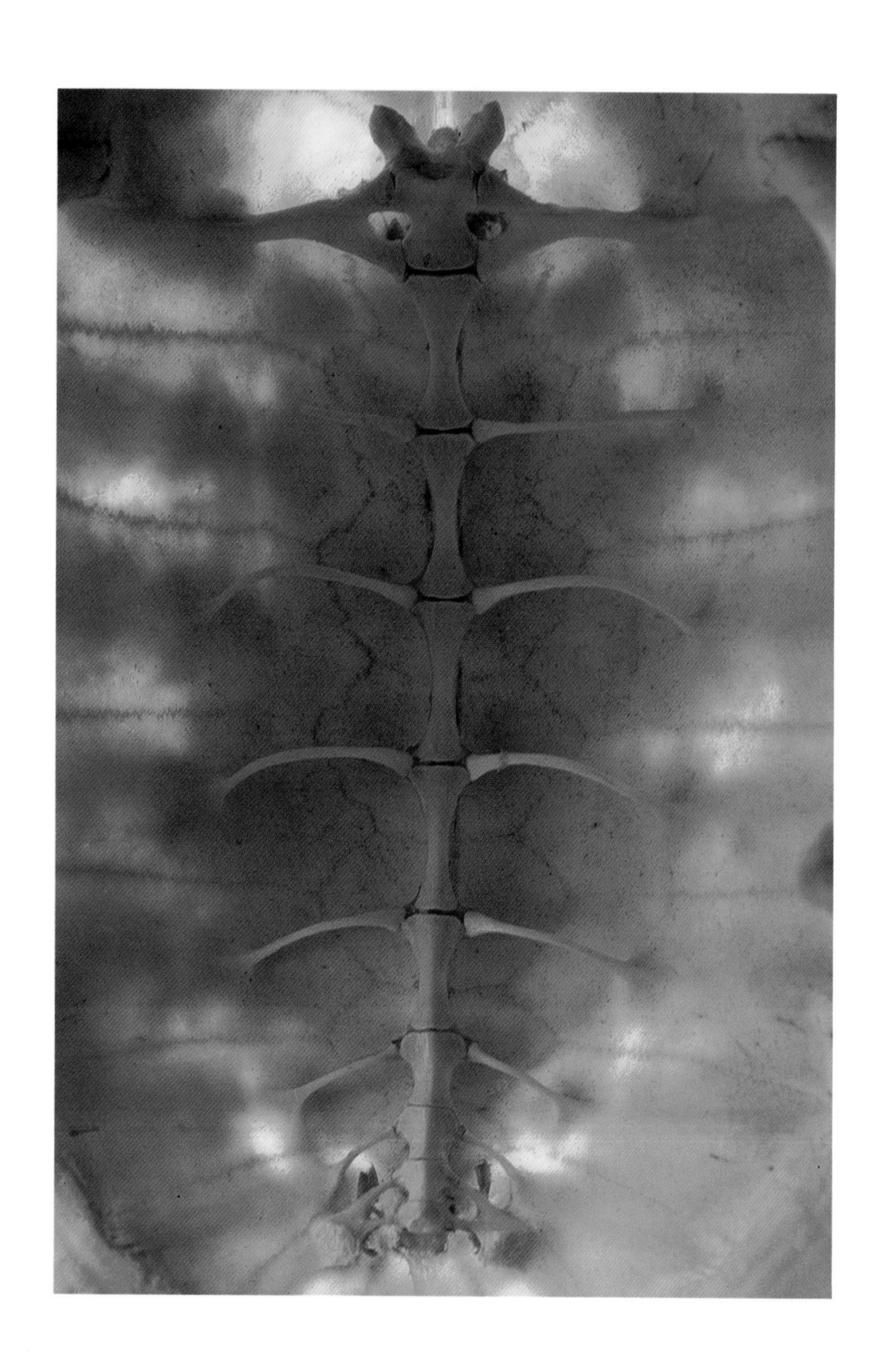

Biology has had its dreamers and visionaries. An illusory search for unity in the design of all organisms has powered the most forthright and uncompromising of these visions. Goethe, the great naturalist (and poet), tried to understand all parts of plants as modified leaves. Etienne Geoffroy Saint-Hilaire, the early nineteenth-century leader of French transcendental biology, sought a common basis of animal form in the vertebra. His noble vision foundered on the false homologies forced by his system. He identified the external skeleton of insects with the internal bones of vertebrates, interpreted each insect segment as a disc of the vertebrate spine, and actually argued that insects dwelled within their own vertebrae. He also regarded the ribs of vertebrates, attached in pairs to the spine, as the legs of arthropods, articulated two by two with the segments. He was wrong, though this smiling turtle might provoke a reassessment.

Or was he? There are things in biology even more basic than form. A short segment of DNA, only 180 base pairs long (and coding therefore for a protein of 60 amino acids), has been found within many of the genes that regulate the content of body segments in Drosophila, the fruit fly. The same stretch of DNA, dubbed the homeobox, has also been discovered in toads, mice, and humans. We may yet find a deeper unity in the genesis of form then Geoffroy ever imagined.

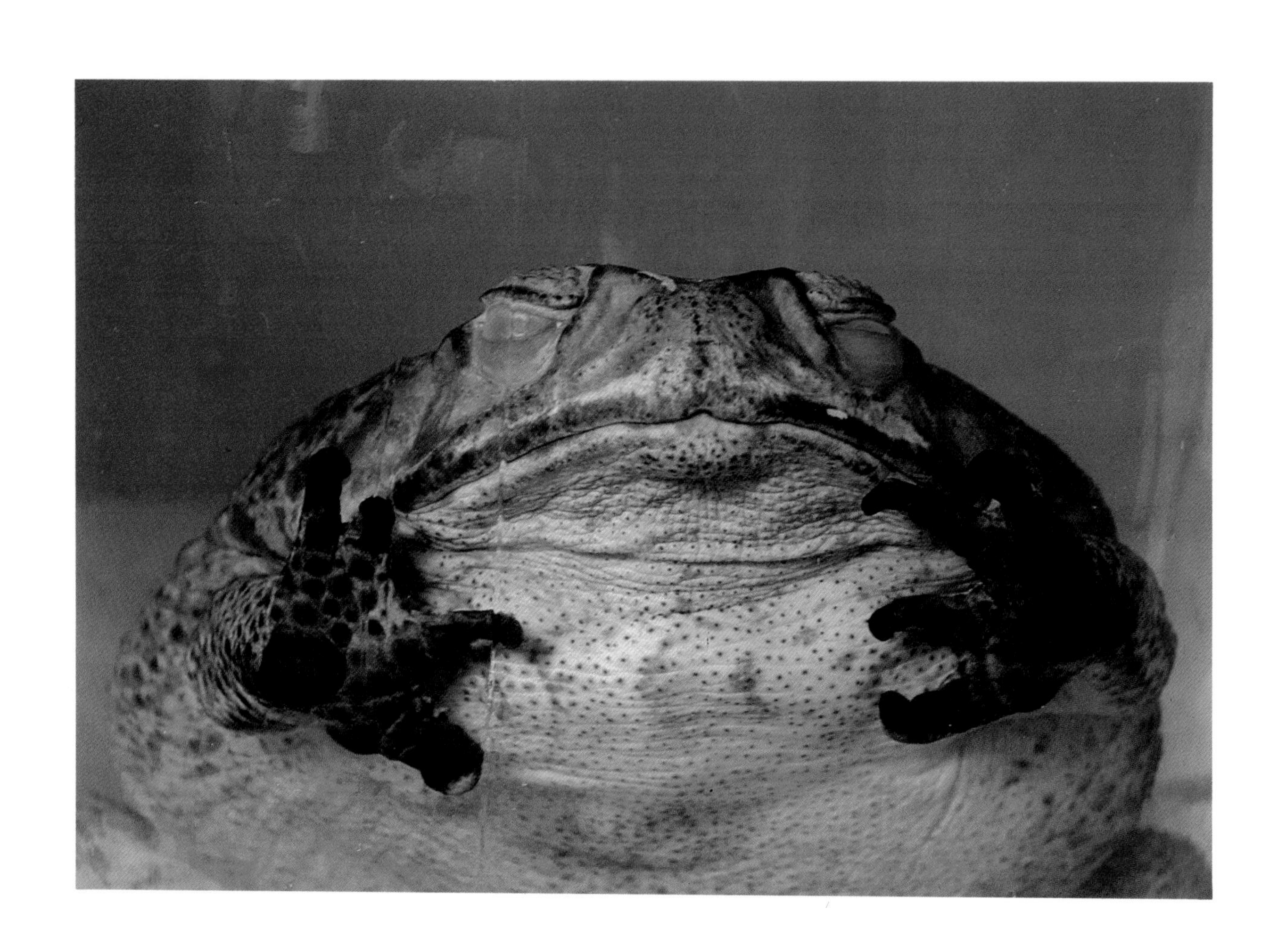

96 TOAD / *Bufo marinus*

UNDERWING MOTH / *Catacala*

Art and science meet here in the most palpable way. The intricate beauty of detailed preservation in this fossil shrimp, some 150 million years old, speaks absolutely for itself.

Such exquisite preservation, including soft parts, is surpassingly rare in the fossil record—so much so that the few known localities provide priceless "windows" into the true diversity of life in ages past (for other times, we have only the coverings of shells or the framework of bones). These "windows" are so rare because they require the unusual conjunction of burial in sediments devoid of oxygen (to prevent decay) and so uniformly fine grained that organic details will not be lost by crushing between the grains of matrix.

These conditions regulated the formation of the famous lithographic limestone of Solnhofen, Germany—home not only of this shrimp, but of the most famous fossil of all: *Archaeopteryx,* the nearly perfect transition between reptile and bird. But the very fine and uniform grain that made such preservation possible has also set the fame and value of Solnhofen for artists. From the invention of the technique to our own day, artists have used stones from the Solnhofen quarries to make all the world's great lithographs, including (to complete the recursion of science to art to science) all the best illustrations of organisms.

Udorella / *Udorella agassizi*

100 VIPER / *Lachesis muta*

We see ecology—nature's transient interactions among species—frozen to quasi-eternity by the confines of a museum storage jar. A viper still seems bent on pursuit of what was, in actuality, its last meal, slit from its stomach after death. The jaws and bellies of snakes are enormously flexible, while the slow digestion of cold-blooded creatures permits an unusual approach to feeding—take something very big once in a very great while. A python may constrict a pig or an antelope (a small human might serve as well), swallow it whole and then retreat to a quiet place and slowly digest for a month or two.

Animals in nature, contrary to the suspicions of cynics or the hopes of idealists, are neither intrinsically vicious nor altruistic. Competition and cooperation are both nature's ways, as the great Darwinian game of struggle for reproductive success dictates. The "struggle for existence" is a general metaphor, but many examples, as here, are ultimate battles about life and death. As Darwin wrote:

> We behold the face of nature bright with gladness, we often see superabundance of food; we do not see, or we forget, that the birds which are idly singing round us mostly live on insects or seeds, and are thus constantly destroying life; or we forget how largely these songsters, or their eggs, or their nestlings, are destroyed by birds and beasts of prey; . . .

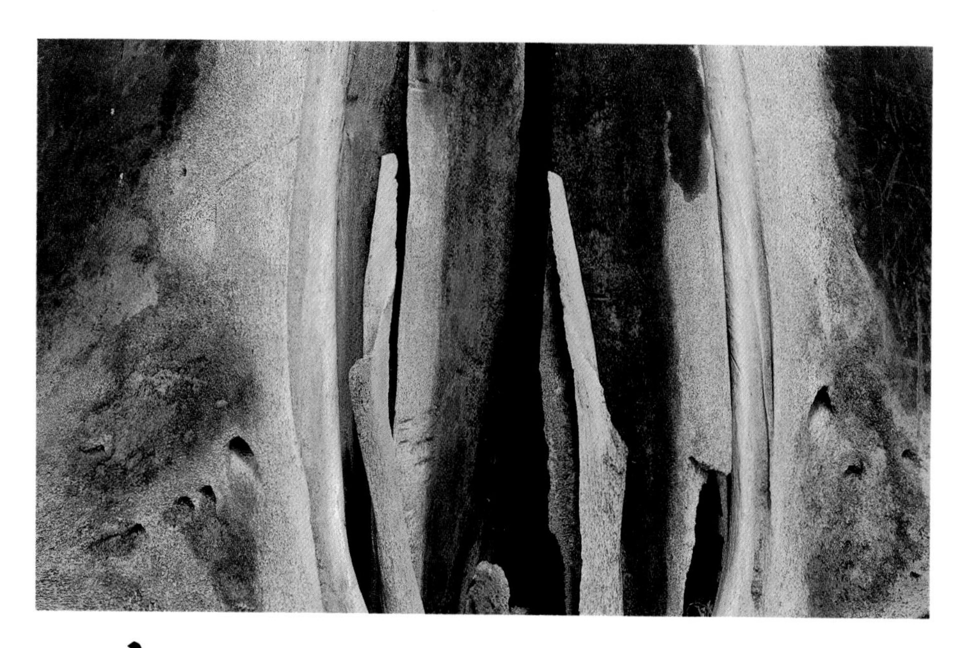

When I understood science far less well than I do today, I used to get annoyed at one of the most common events of professional talks. The speaker would show a slide, often quite beautiful, of some natural object and someone would interrupt by shouting: "What's the scale?" "How crass and narrow-minded," I used to say to myself. "Can't they think of anything more interesting or more conceptual to ask." Now I understand the deep sense behind such a question. The forms of nature are limited and occur at all scales. Nothing is more disconcerting than a picture of something unfamiliar without reference points—for the basic geometries can appear at any size; and we may not know whether we view a photograph of a fly's eye, or an aerial snapshot of a geodesic dome. This ambiguity may destroy the requisite precision of a professional talk, but it may embody the beauty and thought-provoking character of an artist's perception—where a conscious failure to provide scale may accentuate the universality of nature's forms. Here we see, first, the narrow bones of a whale's jaw, and, then, the channels and exit points for nerves and blood vessels etched on the jaw's inner surface. Yet, for color and form alone, we might be viewing America's great southwestern desert at 30,000 feet over Arizona.

WHALE / *Balaenoptera edeni*

If good things come in small packages, why not illustrate this cliché with the biggest animal of all, a whale. We look at an ear region, highlighted by the inflated tympanic bulla. Within this receptacle lie the three smallest bones of all—the sound transmitters of the middle ear, named malleus (hammer), incus (anvil), and stapes (stirrup).

Bones, like words, carry their own history—the primary proof, by the way, of evolution itself. Quirky shifts are as common as predictable transformations. Two of these earbones (hammer and anvil) articulated the jaws of reptilian ancestors—for only the stirrup transmits sound in a lizard's ear. But the stirrup suspended upper jaw to cranium in the jawed fish that gave rise to reptiles. Finally, the precursors of all these bones were parts of gill arches in primordial jawless fishes. The history of life is a labyrinth, not a ladder of progress.

In "Design," Robert Frost mused upon randomness and order in nature by sketching a scene of death that brought three organisms together (a spider eating a moth on a flower). The three have shapes that could not be more different—blob, froth, and sheet.

> A snow-drop spider, a flower like a froth,
> And dead wings carried like a paper kite.

But all are white, an unusual color for each creature. What symbols for order and chaos can we find in this maximal variety of form combined with fortuitous conjunction of color?

We find here the same themes of form and color, as the two archetypal symmetries of invertebrate life meet in mutual white. The crab, like us, is bilaterally symmetrical. The snail is an ever-expanding logarithmic spiral. But the snail's symmetry lies hidden beneath its unique habit. For this is *Xenophora,* the "strange bearer." At regular intervals in its logarithmic expansion, *Xenophora* cements a foreign object to its shell (presumably gaining protection through camouflage). These objects, of different shape and orientation, disrupt the strict symmetry of the shell, although the regular spacing of cementation points forms a record of underlying order in growth. I found it fascinating that the appearance of disorder—the presumed function of this strange cementing habit—should arise as an expression of the shell's basic symmetry.

Xenophora may be quite specific in its choice of adornment. Some select only pebbles, others only shell fragments (aficionados call them "geologists" and "conchologists" respectively).

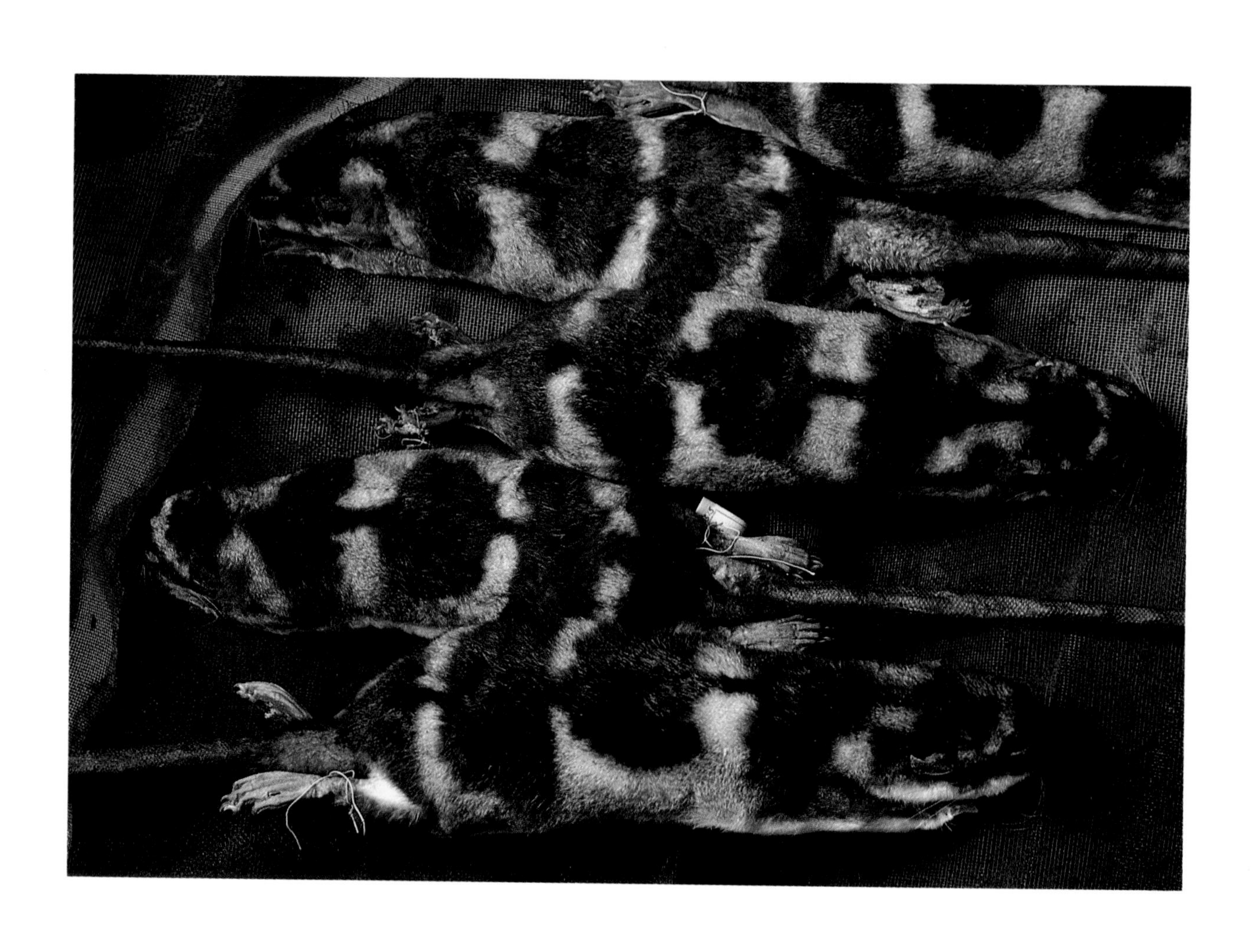

108 YAPOK / *Chironectes minimus*

The Yapok is a fitting symbol of unappreciated diversity. We chauvinistic norteamericanos think of *the* opossum as a single thing and our very own. We call this species the "Virginia opossum" and regard it as a unique "living fossil." But "our" opossum is just one species of a much larger group centered in South America. It is the only marsupial that managed to migrate so far north after the Isthmus of Panama rose just a few million years ago. Yapoks symbolize the diversity of opossums as the most divergent member of the group. They are the only marsupials adapted to aquatic life—with streamlined body, webbed hind feet, and even a waterproof pouch!

We see juxtaposed, the most evident manifestations (in zebras) of a basic principle in organic architecture: serial repetition (an apparently superficial exterior "paint" and the fundamental internal building blocks in this case). Among the three living species of zebras, we find little variation in number of vertebrae but a great range in stripes—from twenty-five to thirty in Burchell's zebra to some eighty in Grevy's zebra, pictured here. A simple unity in construction may underlie this diversity of result. If zebra embryos lay down the stripes 0.4 mm (or about twenty cell diameters) apart, then all diversity of form and number may only reflect the age (and size) of the embryo at first striping. Stripes appear in Burchell's zebra when the embryo is tiny and provides little room for stripes of constant spacing. The larger number and more nearly parallel arrangement of stripes in Grevy's zebra, shown here, may record initial deposition on a larger embryo, a few weeks older and already close enough to final shape that stripes are not bent (as in the other species) into broad sweeping arcs on the rear by an early surge of embryonic growth in zebra hind quarters—a surge that occurs *before* stripes appear in Grevy's zebra, but *after* in the other species.

Contrary to what most Western schoolchildren learn, zebras seem to be black animals with white stripes, as most African people have long believed—an interesting commentary on cultural bias.

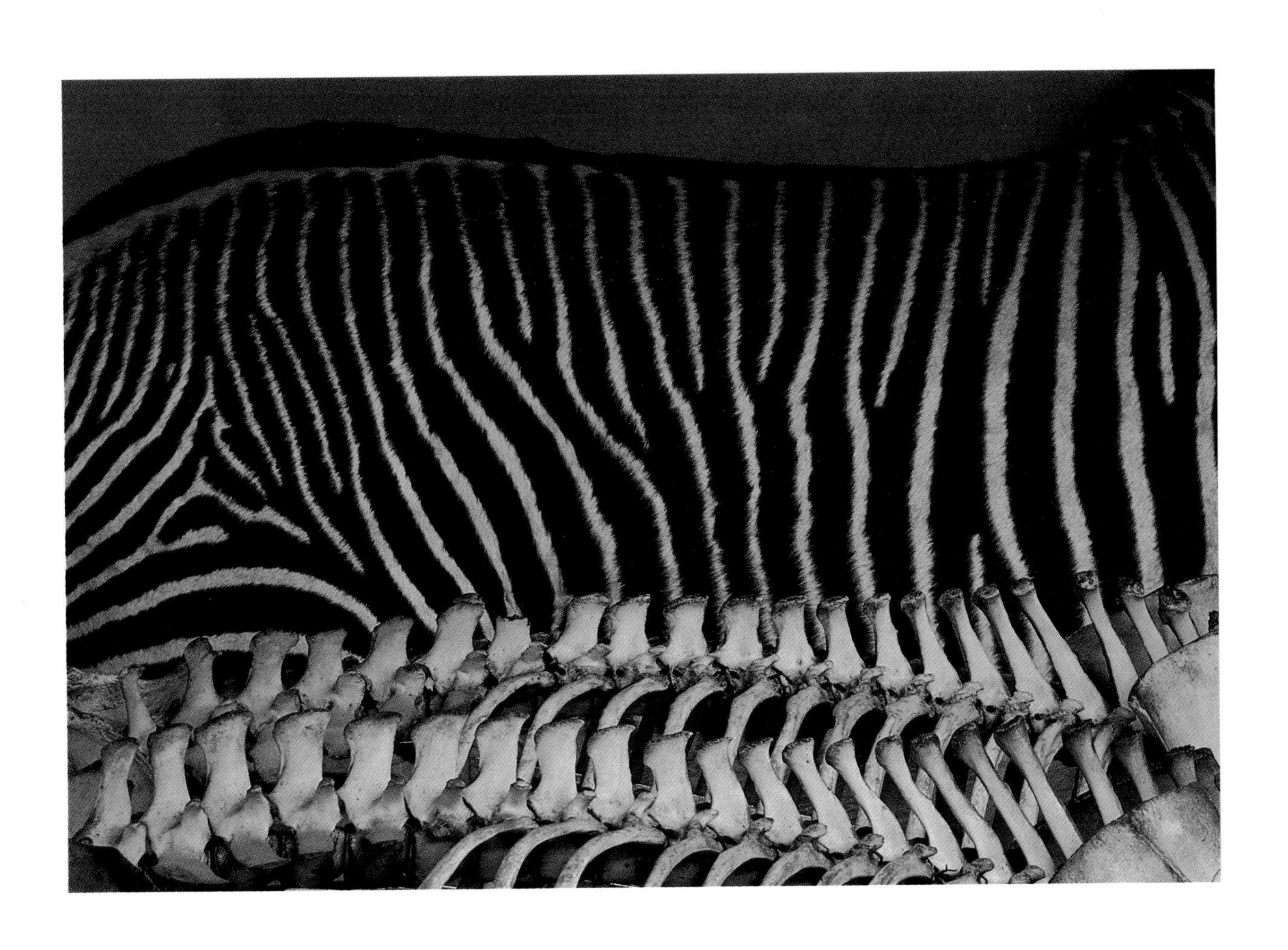

Zebra / *Equus grevyi*

Crabs are our primary cultural symbol of disorder in the sinister sense. We expect elongation and forward motion among the bilaterally symmetrical "higher" creatures—but crabs are rounded and move sidewards. In Aesop's fable, a mother crab castigates her youngster for its ungraceful walk and receives an appropriate reprimand—how thou goest, I will go. Galen tells us that the claw-like extensions of spreading cancers prompted his countrymen to appropriate the animal as a name for this disease of disordered growth—an association sufficiently unpleasant that many newspapers now bowdlerize their astrological charts and relabel the ancient constellation of the crab (Cancer) with the inappropriate "moon children."

Seen from the top, these Japanese crabs of the genus *Zozymus* might elicit the conventional feeling of a bumpy, disconcerting irregularity. But crabs are crustaceans, members of the phylum Arthropoda, and the most ordered of complex animals in their construction as a sequence of repeated segments. (Since 80 percent of animal species are arthropods, mostly insects, the basic design surely seems to work). Seen from the bottom, these same crabs shed their cultural liability—legs neatly ordered in pairs, culminating in frontal claws, the two rows separated by a posterior extension neatly tucked under (stretch it out and crabs proclaim their affinity with elongated relatives among lobsters and shrimp).

Afterword

We began working on this bestiary a year ago. Five years ago, my friends would have testified that my attitude toward natural history specimens and toward the institutions in which they were displayed was one of decided repugnance. A desire to contemplate expressive materials outside of the realm of purely human portraiture, as well as the proximity of the Museum of Comparative Zoology at Harvard, led me to overcome some of this aversion. It is amazing how the light of day falling on these animals can banish feelings of squeamishness and fear. Whether photographing a fossil tooth, a desiccated monkey, or a Bog Woman, I feel a sense of privilege and responsibility.

As a photographer I am attracted to zoological collections by virtue of their fragmentary state. Partially eroded or effaced surfaces appeal to me, as would an ancient piece of fresco, a piece of Egyptian linen with faint hallmarks, a piece of text eaten by termites. In these collections, ranging in time between 1650 and 1979, one can find effects parallel to the visual delights inherent in the worn surfaces of human artifacts.

Although I did honor a request from Steve to consider Cerion, chose the flamingos with his fans in mind, and kept half an eye on the alphabet, the creatures you see were chosen for their visual eloquence and metaphorical suggestiveness. The text, written after the photographs, offers the readers and the artist a different set of metaphors.

Once in a museum drawer, I found the skull of a manatee with a door carved in its jaw which, when swung back, revealed the dental structure above and below the gum-line of the creature. The door was fastened with an ornate brass hinge—for the scientist, a useful device; to the artist, a surrealistic sculpture.

In the domains of the scientist and the curator, the death of an organism is the beginning; it strips away the layer of life. Other layers of information may then vanish with each stage of dissection—the skin from the bone; the bones from the skin; color or opacity from the tissue (see the translucent alizarin-stained frogs, fish, and night monkey); or fluids from the vital organs, arteries, and veins. Irrelevant layers are discarded as the scientist seeks a proper vantage point for his work—be it a study of the vascular system of a squirrel monkey, an analysis of the cavities of an elephant skull, a taxidermic version of a three-toed sloth (with internal padding added to reanimate the skin for the pleasure of the museum-going public). What the viewer to the collections sees then is always partial, sometimes vestigial, and to the nonscientist, often mysterious.

Just as the Bog Woman, by virtue of her privileged status as a human being, had clothes and a personal cultural past,* so each creature herein has a past—a pre-research history that human beings have fabricated out of superstition, and, no doubt, psychological necessity.

"Animals first entered the imagination as messages and promises" (John Berger).** We have rendered them as gods, as totems, as auguries. We have devised peculiar rites for these creatures in natural history museums—inscriptions in ink on bone, chemical baths to render them translucent, systematic placement on shelves, often in the dark, often in shrouds of dust or moth crystals. I think of these treatments as forms of burial, but I think of the animals as expressing, in various ways, life after death.

I tried to use backgrounds or settings that would normally accompany the specimens (the cetacean-oil-stained paper under the dolphin jaws, the cotton batting with the mastadon tooth, the cotton with the squirrel monkeys, the glass lid on the box of ibis eggs). No specimen in liquid was photographed out of its liquid—the fish on the parchment music remains in glycerine as does the mouse-tailed bat, the night monkey, and the frogs. The viper (bushmaster) and bird are, as they were, in the same bottle in alcohol, the angler fish also in alcohol. In some cases I have used simple alien backgrounds, either of metal or paper. While I have transcribed the Latin names from the original tag or label accompanying each specimen, these names may not reflect current taxonomic usage.

With the exception of two constructed photographs (the capuchin in the box and the macaque with the ball), and the zebra and the elephant located in gloomy cabinets under artificial lights, all subjects were photographed in natural light alone. Whenever possible I used sunlight.

Not unlike the scientist who manipulates his specimens to suit his subjective ends, I occasionally exercise instinctive fictional muscles to convey more than the "facts." The creatures in the introduction (with the exception of the horse's jaw) are confined—either in "cages" or, as in marginalia of the medieval bestiaries, to the page. Here I hint at an immense mythical world also of the beast—a world of anomalies and monsters, both of which hovered at the brink, but not in the midst, of this volume.

With the exception of the angler fish at the National Museum of Natural History, Dublin, none of these animals are currently on display; they are in the domain of the research scientist. To many curators of scientific collections, a browser at the door, with a tripod and without calipers, tends, upon occasion, to raise eyebrows, not to mention blood pressure. This book is dedicated to all my curator friends and foes, the first for their trusting forbearance, the second for whatever distress I imposed upon them in an effort to gain access to materials for the aesthetic thrill of it. Clearly there is room for far more dialogue between my discipline and theirs.

*P. V. Glob. *The Bog People.* Ithaca, N.Y.: Cornell University Press, 1969 (the Huldremose Woman).

**John Berger. *About Looking.* New York: Pantheon, 1980, Chapter 1—Why Look at Animals?

Technical note: The majority of these photographs were taken originally as 35mm transparencies, on a Nikkormat EL, most frequent film Kodachrome 25, most frequent lens 55mm; the longest exposures were 3–4 minutes for the whale jaws. The monkey with the ball and the monkey in the box are 20 x 24 Polaroid Land Prints.

Photographs in the Introduction include two alizarin-stained fish and an alizarin-stained shark, fish scales mounted on paper accompanied by a handwritten description by a nineteenth-century collector, four *Xenophora* placed on a termite-eaten text, illustrations from the *Museum Wormarium Lugduni Batavorum,* Copenhagen, 1655, and photographs of a horse's jaw embedded in a tree from the Zoologisk Museum, Copenhagen.

Acknowledgments

In the bestiary, the Angler Fish comes from the National Museum of Natural History, Dublin; the Colugos, Cat, Elephant, Fishes, Kiwis, Proboscis Monkeys, Sakis, Whales, *Xenophora,* Zebra, Zozymus Crabs come from the Rijksmuseum van Natuurlijke Historie, Leiden; the Squirrel Monkeys, Grampus, and Night Monkey come from the Smithsonian Museum of Natural History, Washington; the Sloth, from the Zoologisk Museum, Copenhagen; and the Bog Woman from the National Museum, Copenhagen. The remaining specimens all come from the Museum of Comparative Zoology, Harvard University, Cambridge.

A fair number of photographs were done at the Rijksmuseum van Natuurlijke Historie in Leiden. I extend my thanks to the generous staff of that august collection—Dr. Lipke B. Holthuis, Dr. G. V. F. Mees, Martien J. P. van Oijen, and Dr. Chris Smeenk, who entertained my interruptions and queries with unfailing good spirit, patience, and energy. In Copenhagen, I thank Dr. Nana Noe-Nygaard, Dr. Svend Erik Bedix-Almgreen of the Geologisk Museum, Jorgen Jensen of the National Museum, and Dr. Hans Baagøe of the Zoology Museum of the University of Copenhagen. In Dublin, I thank Nigel Monahan and Grace Griffiths of the National Museum. In Washington, I thank Charles Potter and Dr. Richard Thorington of the Smithsonian Museum of Natural History. In Boston, I thank the Boston Athenaeum for allowing the photographs from Ole Worm's famous catalog. In Cambridge, I thank Una MacDowell and Kathleen Scully of the Peabody Museum.

Collection managers and curators in virtually every department of the Museum of Comparative Zoology at Harvard University were helpful to this project. I am grateful to Dr. Pere Alberch, Ron Eng, Dr. Stephen Jay Gould, Artis Johnston, Dr. Raymond Paynter, Jr., Jose Rosado, Chuck Shaff, Scott Shaw, Dr. Melanie Stiassny, and Charlie Vogt. Many thanks to Mrs. Agnes Pilot for her help in preparing the manuscript. Many thanks to the Director of the Museum, Dr. James McCarthy, for his support.

Looking back to the origins of my infiltration into the Museum of Comparative Zoology, I wish to thank Gabrielle Dundon, Director of Public Programs, and Ed Haack, of the Exhibits Department. Dr. Charles Lyman and Dr. John

Kirsch treated my arrival in the Mammal Department with courtesy and friendship. I am most especially indebted to the staff of the Mammal Department—to Maria Rutzmoser, Marilyn Massaro, and Jane Winchell, and to Professor Rodney Honeycutt, for lending me more than specimens—tabletops in sunny corners, coffee cups, timely suggestions about likely candidates for this project, and above all, their encouragement and friendship.

I am very grateful to Jim Rohan and Carmine Dorato of Positive Photographics who worked long hours to produce splendid reproductions of my original photographs.

Thanks and love to those friends who traveled with me in fact and spirit for this book: Laurie Godfrey, Paul Levenson, Wendy MacNeil, Marcuse Pfeifer, Karen Reuter, and Sam Yanes.

And love to the forebearing fellows at home—Dennis, Andrew, and John Henry Purcell.